Series / Number 07-079

REGRESSION DIAGNOSTICS

JOHN FOX
York University

SAGE PUBLICATIONS
The International Professional Publishers
Newbury Park London New Delhi

To the memory of my father, Joseph, and for my son, Jesse Joseph.

For information address:

SAGE Publications, Inc.
2455 Teller Road
Newbury Park, California 91320

SAGE Publications Ltd.
6 Bonhill Street
London EC2A 4PU
United Kingdom

SAGE Publications India Pvt. Ltd.
M-32 Market
Greater Kailash I
New Delhi 110 048 India

Printed in the United States of America

Fox, John. 1947-
Regression diagnostics/ John Fox.
p. cm.—(A Sage university papers series. Quantitative applications in the social sciences; 79)
Includes bibliographical references.
ISBN 0-8039-3971-X
1. Regression analysis. 2. Social sciences—Statistical methods.
I. Title. II. Series: Sage university papers series. Quantitative applications in the social sciences; no. 79.
QA278.2.F63 1991
519.5'36—dc20 91-24057

92 93 94 10 9 8 7 6 5 4 3 2

Sage Production Editor: Diane S. Foster

When citing a university paper, please use the proper form. Remember to cite the current Sage University Paper series title and include the paper number. One of the following formats can be adapted (depending on the style manual used):
(1) FOX, J. (1991) Regression Diagnostics: An Introduction. Sage University Paper Series on Quantitative Applications in the Social Sciences, 07-079. Newbury Park, CA: Sage.
OR
(2) Fox, J. (1991). *Regression Diagnostics: An Introduction* (Sage University Paper series on Quantitative Applications in the Social Sciences, series no. 07-079). Newbury Park, CA: Sage.

CONTENTS

SERIES EDITOR'S INTRODUCTION

In social science data analysis, no technique receives more use than regression. With modern interactive personal computers, getting an estimated regression equation is as easy as 1-2-3, quite literally, because a researcher with any commonly available software program can operate as follows: 1—load the sample data; 2—specify the equation; 3—estimate with ordinary least squares, which might yield something like this:

$$Y = 62 + 71.5X_1 + 5.4X_2 + e$$

Do these estimates tell us how the world really works? For example, does a unit change in X_1 produce an expected change of 71.5 in Y, given that X_2 is held constant? We would like to speak with confidence about the accuracy of that population estimate. However, our faith in the regression result depends on coping successfully with common problems: multicollinearity, outliers, non-normality, heteroscedasticity, and nonlinearity.

Professor Fox explicates "diagnostics" for uncovering these problems. Take, for example, the problem of outlying observations, or of influential observations generally. In addition to the usual graphic methods, which may illustrate how one extreme observation can "leverage" the regression line, Fox also interprets many other measures: *hat-values, studentized residuals, Cook's* D, and *partial-regression plots*. Conveniently, most of these measures are routinely available on well known software programs, such as SAS or SPSS.

After diagnosis, Fox carefully considers possible solutions. Here is a sample of the issues: With high multicollinearity, should a variable be removed? If there is an outlier, should it be discarded? When the error distribution is skewed, should a transformation be applied? Given heteroscedasticity, ought a weighted-least-squares solution be performed? In the face of nonlinearity, is a power transformation in order? In answering such important questions, mechanical quick-fixes are avoided. As the author emphasizes repeatedly throughout, the methods are "no substitute for judgment and thought."

To enrich his explanations, Fox draws on a variety of data examples: the U.S. Census count, occupational prestige, reported body weights, and interlocking directorates among Canadian firms. The examples make the diagnostics accessible to a broad range of regression users. Further, for those interested in more advanced treatments, much is offered in his technical appendices (e.g., an evaluation of ridge regression as a solution to high multicollinearity). Everyone who carries out a regression analysis should, as a matter of course, apply a battery of diagnostics. There is no better introduction to regression diagnostics than this monograph.

—*Michael S. Lewis-Beck*
Series Editor

ACKNOWLEDGMENTS

Many individuals provided me with useful comments on drafts of this monograph: Ken Bollen, Gene Denzel, Mike Friendly, Lydia Kiselyk, Scott Long, Georges Monette (and his students), Bob Stine, Yu Xie, the editor of this series, two anonymous referees, and the participants in the 1990 ICPSR workshop on regression diagnostics. If I haven't profited from all of this help, then the fault surely is mine. I am grateful to Caroline Davis; Michael Ornstein; Bernard Blishen, William Carroll, and Catherine Moore; the Inter-University Consortium for Political and Social Research; and the National Opinion Research Center for providing me with data for several examples, and to Eugene Ericksen for responding to questions about his work on the census undercount. Of course, none of these individuals or institutions is responsible for the use to which I have put their data. I am also grateful to the several publishers who granted permission to redraw or reprint copyrighted materials. Finally, I would like to thank the Population Studies Center of the University of Michigan, and particularly its past director Bill Mason, for providing a stimulating atmosphere during my sabbatical year when I drafted this monograph, and to acknowledge the support provided by York University and the Social Sciences and Humanities Research Council of Canada for my sabbatical research.

REGRESSION DIAGNOSTICS
An Introduction

John Fox
York University

1. INTRODUCTION

Linear least-squares regression analysis is the most widely used statistical technique in social research and provides the basis for many other statistical methods. Yet least-squares regression is susceptible to a variety of difficulties and makes strong and often unreasonable assumptions about the structure of the data. *Regression diagnostics* are techniques for exploring problems that compromise a regression analysis and for determining whether certain assumptions appear reasonable.

Although examination of data for potential difficulties has always been a hallmark of good data analysis, the modern development of regression diagnostics coincides with the ready availability of computers for interactive statistical analysis and is thus a product largely of the past 2 decades. Closely associated with methods of regression diagnostics are techniques for correcting problems that are detected. Many of these methods employ transformations of the data.

As a preliminary example, consider the four scatterplots from Anscombe (1973) shown in Figure 1.1. One of the goals of statistical analysis is to provide an adequate descriptive summary of the data. All four of Anscombe's data sets were contrived cleverly to produce the same standard linear-regression outputs: slope, intercept, correlation, regression standard error, coefficient standard errors, and statistical tests—but, importantly, not the same residuals.

In Figure 1.1a, the linear regression is a reasonable description of the tendency for y to increase with x. In Figure 1.1b, the linear regression fails to capture the obviously curvilinear pattern of the data—the linear model is clearly wrong. In Figure 1.1c, one data point is out of line with the others and has an undue influence on the fitted regression line. A line through the other points fits them perfectly. Ideally in this case, we want to understand why the last observation differs

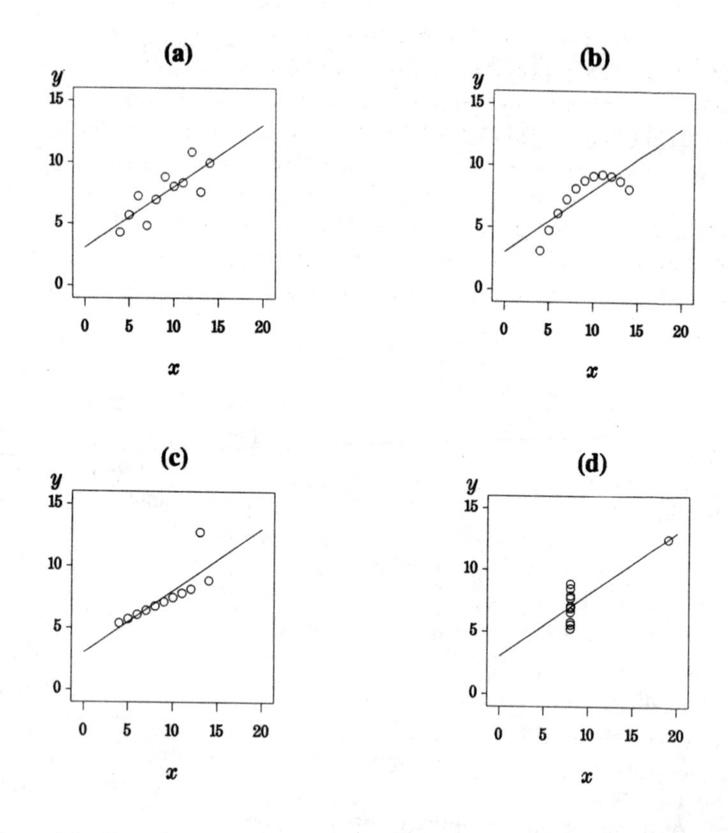

Figure 1.1. Four data sets with identical standard regression outputs, from F.J. Anscombe (1973). The least-squares regression line is shown on each scatterplot.

SOURCE: Anscombe (1973). Redrawn and reprinted with the permission of the American Statistical Association.

from the others—possibly it is special in some way (e.g., it is strongly affected by a variable other than x), or represents an error in recording the data. Of course, we are exercising our imaginations here, because Anscombe's data are simply made up, but the essential point is that we need to address anomalous data substantively. In Figure 1.1d, in contrast, we are unable to fit a line at all but for the last data point. At the very least, we should be reluctant to trust the estimated regression coefficients because of their dependence on this one point.

Anscombe's simple illustrations serve to introduce several of the themes of this monograph, including nonlinearity, outlying data, influential data, and the effectiveness of graphical displays. The usual regression outputs clearly do not tell the whole story. Diagnostic methods—many of them graphical—help to fill in the missing parts.

The monograph begins in Chapter 2 with a review of least-squares linear regression. Chapter 3 takes up the problem of collinearity in multiple regression. Chapter 4 deals with outlying and influential data. Chapters 5, 6, and 7 take up non-normality of errors, nonconstant error variance, and nonlinearity. Chapter 8 briefly considers problems and opportunities presented by discrete data. Chapter 9 introduces sophisticated diagnostics based on maximum-likelihood methods, score tests, and constructed variables. In Chapter 10, I step back from specific diagnostic methods to suggest how these techniques can be applied effectively in research. This chapter includes some remarks on implementing diagnostics with standard statistical computer packages and concludes with recommendations for further reading.

Much technical detail has been relegated to the Appendix, the parts of which are keyed to specific sections of the text. The primary prerequisites for understanding the material in the Appendix are acquaintance with the matrix algebra of least-squares regression and elementary statistical theory; and though the Appendix provides more depth to the presentation, it may be skipped by the less technically inclined reader. My goal is to make this monograph relatively self-contained, while maintaining its general accessibility.

The monograph is far from exhaustive, but I have tried to deal with the central issues in validating regression models. Primarily because of space considerations, however, there is no treatment of autocorrelated errors in time-series regression, except for a brief mention in Chapter 10. The topic also is covered in detail in another monograph in this series (Ostrom, 1990)

2. LINEAR LEAST-SQUARES REGRESSION

I assume that the reader is familiar with linear-regression analysis, and consequently this chapter serves primarily as a review and to establish notation. A more technical treatment appears in Appendix A2.1.

The Regression Model

The linear-regression model is given by the equation

$$y_i = \beta_0 + \beta_1 x_{1i} + \beta_2 x_{2i} + \ldots + \beta_k x_{ki} + \varepsilon_i \quad [2.1]$$

for $i = 1, \ldots, n$ sampled observations. In Equation 2.1, y_i is the dependent variable, the x_{ji} are regressors, and ε_i is an unobservable error. The β_j are unknown parameters to be estimated from the data. It is standard to assume that the errors are normally and independently distributed with zero expectation (mean) and constant variance σ^2: $\varepsilon_i \sim \text{NID}(0, \sigma^2)$. Consequences of violating these assumptions, and methods for detecting violations, are considered later in the monograph.

If the x_{ji} are sampled along with the y_i, rather than fixed by design as in an experiment, then it is additionally assumed that the *x*s are distributed independently of the εs. This last assumption may be construed either descriptively or structurally. Descriptively, the mean *y* value in the population for any combination of *x* values must lie on the regression surface. Structurally, or causally, we require in addition that omitted causes of *y* (which are thus components of the error) that are not themselves effects of the *x*s are linearly uncorrelated with the *x*s. Except under special circumstances, this last assumption cannot be checked from the data, because the least-squares fit ensures that the residuals, which estimate the errors, are uncorrelated with the *x*s in the sample.

Least-Squares Estimation

The fitted regression model is written

$$y_i = b_0 + b_1 x_{1i} + b_2 x_{2i} + \ldots + b_k x_{ki} + e_i = \hat{y}_i + e_i$$

where y_i and the x_{ji} are as before, the b_j are estimates of the corresponding β_j, and the e_i are residuals. The fitted values are given by $\hat{y}_i = b_0 + b_1 x_{1i} + \ldots + b_k x_{ki}$. The least-squares regression coefficients, chosen so as to minimize the sum of squared residuals, are the values of the b_j that satisfy the normal (or estimating) equations:

$$
\begin{aligned}
b_0 n + b_1 \Sigma x_1 \quad + \ldots + b_k \Sigma x_k \quad &= \Sigma y \\
b_0 \Sigma x_1 + b_1 \Sigma x_1^2 \quad + \ldots + b_k \Sigma x_1 x_k &= \Sigma x_1 y \\
&\vdots \\
b_0 \Sigma x_k + b_1 \Sigma x_1 x_k + \ldots + b_k \Sigma x_k^2 \quad &= \Sigma x_k y
\end{aligned}
$$

Because the sums are obviously over $i = 1, \ldots, n$, I have suppressed the subscript i for observations (e.g., x_1 represents x_{1i}). The normal equations have a unique solution for the b_j as long as (a) none of the x_j is constant, and (b) none of the x_j is a perfect linear function of others.

The normal equations imply that the least-squares residuals sum to zero and thus have a mean of zero. Furthermore, the residuals are uncorrelated with the fitted values and with the xs because

$$\Sigma e_i \hat{y}_i = 0$$

$$\Sigma e_i x_{ji} = 0, \quad j = 1, \ldots, k$$

The error variance is estimated by $s^2 = \Sigma e_i^2/(n - k - 1)$, where $n - k - 1$ is the degrees of freedom for error. The squared multiple correlation for the fitted model, given by

$$R^2 = \frac{\Sigma (y_i - \bar{y})^2 - \Sigma e_i^2}{\Sigma (y_i - \bar{y})^2} = \frac{\Sigma (\hat{y}_i - \bar{y})^2}{\Sigma (y_i - \bar{y})^2}$$

is interpreted as the proportion of variation in y captured by its linear regression on the xs.

Statistical Inference for Regression Coefficients

Estimated coefficient sampling variances for $b_1, \ldots, b_k$ are given by

$$\hat{V}(b_j) = \frac{s^2}{\Sigma (x_{ji} - \bar{x}_j)^2} \times \frac{1}{1 - R_j^2} = \frac{s^2}{(n-1)\, s_j^2} \times \frac{1}{1 - R_j^2}$$

where $s_j^2 = \Sigma(x_{ji} - \bar{x}_j)^2/(n-1)$ is the variance of x_j, and R_j^2 is the squared multiple correlation from the regression of x_j on the other xs. A t statistic for the hypothesis H_0: $\beta_j = \beta_j^{(0)}$ (usually, H_0: $\beta_j = 0$) is given by $t_0 = (b_j - \beta_j^{(0)})/\text{SE}(b_j)$, where $\text{SE}(b_j) = [\hat{V}(b_j)]^{1/2}$ is the estimated standard error of b_j. Under H_0, the test statistic t_0 is distributed as a t variable with $n - k - 1$ degrees of freedom.

To test the hypothesis that a set of regression coefficients (excluding the constant β_0) is zero, for example, H_0: $\beta_1 = \beta_2 = \ldots = \beta_p = 0$ (where $p \leq k$), we calculate the incremental F statistic

$$F_0 = \frac{n-k-1}{p} \times \frac{R^2 - R_0^2}{1 - R^2}$$

Here R^2 is, as before, the squared multiple correlation from the full model, and R_0^2 is the squared multiple correlation for the regression of y on the remaining xs: $x_{p+1}, \ldots, x_k$. If $p = k$, then $R_0^2 = 0$. These t and F tests are exact under the assumptions of the model, including the assumption of normally distributed errors.

A $100(1-\alpha)\%$ confidence interval for β_j is given by

$$\beta_j = b_j \pm t_{\alpha/2,\, n-k-1}\, \text{SE}(b_j) \qquad [2.2]$$

Because the width of the confidence interval is proportional to the estimated coefficient standard error, $\text{SE}(b_j)$ is a natural measure of the precision of the estimator b_j.

Likewise, an ellipsoidal joint confidence region for several coefficients may be constructed from the coefficient variances and covariances along with a critical value from the F distribution (see Appendix A2.1). An illustration for two parameters, β_1 and β_2, appears in Figure 2.1. Just as the confidence interval in Equation 2.2 gives all values of β_j acceptable at level α, the ellipse in Figure 2.1 encloses all *jointly* acceptable values of β_1 and β_2.

The confidence ellipse is centered on the estimates b_1 and b_2. The projections of the ellipse onto the β_1 and β_2 axes give individual confidence intervals for these parameters, though at a somewhat higher level of confidence than the joint region. Just as the length of a confidence interval expresses the precision of estimation of a single coefficient, the size of a joint confidence region for several coefficients (i.e., area for two βs, volume for three, and

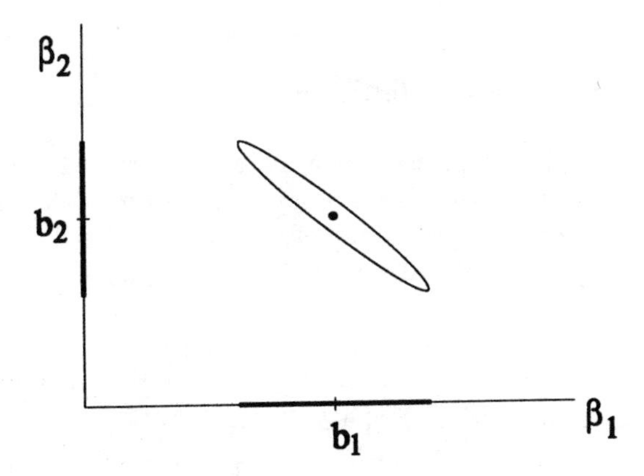

Figure 2.1. The joint confidence region for two regression coefficients, β_1 and β_2. The confidence ellipse is centered at the estimates, b_1 and b_2. The projection of the joint confidence ellipse onto the β_1 and β_2 axes provides individual confidence intervals for these parameters (but at a larger confidence level than the joint confidence region).
SOURCE: Figure courtesy of Georges Monette.

hypervolume for four or more) expresses their simultaneous precision of estimation.

The General Linear Model

Because no distributional assumptions are made about the *x*s, other than uncorrelation with the errors, the domain of the linear-regression model is broader than it first appears. The *x*s may include dummy regressors constructed to capture the effects of a qualitative independent variable; interaction regressors, formed as products of other regressors, for non-additive effects of independent variables; polynomial regressors to model nonlinear patterns in the data; and so on. As long as the model can be expressed in the form of Equation 2.1—that is, as long as the model is linear in the parameters $\beta_0, \beta_1, \ldots, \beta_k$—it can be accommodated by linear-regression analysis: The regression surface itself need not be flat. In this broad context, the linear-regression model is often termed the general linear model.

3. COLLINEARITY

Collinearity and Variance Inflation

As mentioned in Chapter 2, when there is a perfect linear relationship among the regressors in a linear-regression model, the least-squares coefficients are not uniquely defined. This result is easily seen for $k = 2$ xs, for which the normal equations are

$$\begin{aligned} b_0 n + b_1 \Sigma x_1 + b_2 \Sigma x_2 &= \Sigma y \\ b_0 \Sigma x_1 + b_1 \Sigma x_1^2 + b_2 \Sigma x_1 x_2 &= \Sigma x_1 y \\ b_0 \Sigma x_2 + b_1 \Sigma x_1 x_2 + b_2 \Sigma x_2^2 &= \Sigma x_2 y \end{aligned} \qquad [3.1]$$

Solving the normal equations produces

$$\begin{aligned} b_0 &= \bar{y} - b_1 \bar{x}_1 - b_2 \bar{x}_2 \\ b_1 &= \frac{\Sigma x_1' y' \Sigma x_2'^2 - \Sigma x_2' y' \Sigma x_1' x_2'}{\Sigma x_1'^2 \Sigma x_2'^2 - (\Sigma x_1' x_2')^2} \\ b_2 &= \frac{\Sigma x_2' y' \Sigma x_1'^2 - \Sigma x_1' y' \Sigma x_1' x_2'}{\Sigma x_1'^2 \Sigma x_2'^2 - (\Sigma x_1' x_2')^2} \end{aligned} \qquad [3.2]$$

where $x_1' = x_1 - \bar{x}_1$, $x_2' = x_2 - \bar{x}_2$, and $y' = y - \bar{y}$ are variables in mean-deviation form.

The correlation between x_1 and x_2 is given by

$$r_{12} = \frac{\Sigma x_1' x_2'}{\sqrt{\Sigma x_1'^2 \Sigma x_2'^2}}$$

Thus, if $r_{12} = \pm 1$, then the denominator of b_1 and b_2 in Equation 3.2 is zero, and these coefficients are undefined. (More properly, there is an infinity of pairs of values of b_1 and b_2 that satisfy the normal equations [Equation 3.1].)

A strong, but less than perfect, linear relationship between the xs causes the least-squares regression coefficients to be unstable: Coefficient standard errors are large, reflecting the imprecision of estimation of the βs; consequently, confidence intervals for the βs are broad. Small changes in the data—even, in extreme cases, due to rounding errors—can substantially alter the least-squares coefficients, and relatively large changes in the coefficients from the least-squares values hardly increase the sum of squared residuals.

Recall from the previous chapter that the estimated variance of the least-squares regression coefficient b_j is

$$\hat{V}(b_j) = \frac{s^2}{(n-1)\,s_j^2} \times \frac{1}{1 - R_j^2} \qquad [3.3]$$

The impact of collinearity on the precision of estimation is captured by $1/(1 - R_j^2)$, called the variance-inflation factor VIF_j. It is important to keep in mind that it is not the pairwise correlations among the regressors (when $k > 2$) that appears in the VIF, but the multiple correlation for the regression of a particular x on the others. For this reason, collinearity in multiple regression is sometimes termed *multicollinearity*.

Note, incidentally, that the other factors affecting the precision of estimation in Equation 3.3 are the estimated error variance, the sample size, and the variance of x_j. Small error variance, large sample size, and spreadout xs all contribute to precise estimation in regression. It is my experience that imprecise estimates in social research are more frequently the product of large error variance and relatively small samples than of serious collinearity.

Because the precision of estimation of β_j is most naturally expressed as the width of the confidence interval for this parameter, and because the width of the confidence interval is proportional to the standard error of b_j, I recommend examining the square root of the VIF in preference to the VIF itself. Table 3.1 reveals that the linear relationship among the xs must be very strong before collinearity seriously degrades the precision of estimation: For example, it is not until R_j approaches 0.9 that the precision of estimation is halved.

Consider, by way of example, the regression analysis reported in Table 3.2, from data presented by Ericksen, Kadane, and Tukey (1989). The object here was to develop a prediction equation to improve estimates of the 1980 U.S. Census undercount. It is well established that the census fails to count all residents of the country

TABLE 3.1
Coefficient Variance Inflation as a Function of Inter-Regressor Multiple Correlation

R_j	$VIF_j = 1/(1 - R_j^2)$	$\sqrt{VIF}$ [a]
0.0	1.0	1.0
0.2	1.04	1.02
0.4	1.19	1.09
0.6	1.56	1.25
0.8	2.78	1.67
0.9	5.26	2.29
0.95	10.3	3.20
0.99	50.3	7.09
0.999	500.0	22.4
1.0	∞	∞

a. Impact on the standard error of b_j.

and that the likelihood of being missed is greater for certain categories of individuals, such as nonwhites, the poor, and residents of large cities. The dependent variable in the regression is a preliminary estimate of the undercount for each of 66 areas into which the country was divided by the authors. The 66 areas included 16 large cities, the remaining portions of the 16 states in which the cities are located, and the other 34 states. The preliminary estimates are regressed on eight predictors thought to influence the undercount, including

1. the percentage black or Hispanic (Minority);
2. the rate of serious crimes per 1000 population (Crime);
3. the percentage poor (Poverty);
4. the percentage having difficulty speaking or writing English (Language);
5. the percentage age 25 and older who have not finished high school (High school);
6. the percentage of housing in small, multiunit buildings (Housing);
7. a dummy variable coded one for cities, zero for states and state remainders (City); and
8. the percentage of households counted by "conventional" personal enumeration, as opposed to mail-back questionnaire with follow-ups (Conventional).

TABLE 3.2

Regression of Estimated 1980 U.S. Census Undercount on Area Characteristics for 66 Central Cities, State Remainders, and States

Predictor[a]	*Coefficient*	*Standard Error*	$\sqrt{VIF}$
Constant	−1.77	1.38	—
Minority	0.0798	0.0226	2.24
Crime	0.0301	0.0130	1.83
Poverty	−0.178	0.0849	2.11
Language	0.215	0.0922	1.28
High school	0.0613	0.0448	2.15
Housing	−0.0350	0.0246	1.37
City	1.16	0.77	1.88
Conventional	0.0370	0.0093	1.30
R^2	0.708		

SOURCE: Data taken from Ericksen, Kadane, and Tukey (1989).

NOTE: These authors employed a weighted-least-squares regression (cf., Appendix A6.2) to take account of the different precision of initial estimates of the undercount in the 66 areas. The results reported here, in contrast, are for an ordinary-least-squares regression.

Correlations among the predictors appear in Table 3.3. Although some of the pairwise correlations are moderately large—the biggest around 0.75—none is close to one. It is apparent from the square-root VIFs shown in Table 3.2, however, that the precision of several of the regression estimates—in particular, the coefficients for Minority, Poverty, and High school—suffers from collinearity.

Coefficient variance inflation, as a direct index of the extent to which collinearity harms estimation, can be extended to the joint confidence region for several coefficients. Relevant applications include contexts—such as sets of dummy regressors or polynomial regressors—where the variance inflation of individual coefficients in the set is of secondary interest at best. (See Fox and Monette [forthcoming] for details.)

Coping with Collinearity: No Quick Fix

When collinearity between x_1 and x_2 is strong, for example, the data contain little information about the impact of x_1 on y holding x_2 constant statistically, because there is little variation in x_1 when x_2 is

TABLE 3.3
Correlations Among Eight Predictors of the 1980 Census Undercount

Predictor	*Minority*	*Crime*	*Poverty*	*Language*	*High School*	*Housing*	*City*
Crime	0.655						
Poverty	0.738	0.369					
Language	0.395	0.512	0.152				
High school	0.535	0.0666	0.751	–0.116			
Housing	0.356	0.532	0.335	0.340	0.235		
City	0.758	0.729	0.538	0.480	0.315	0.566	
Conventional	–0.334	–0.233	–0.157	–0.108	–0.414	–0.0863	–0.269

SOURCE: Data taken from Ericksen, Kadane, and Tukey (1989).

fixed. Of course, the same is true for x_2 fixing x_1. Because b_1 estimates the partial effect of x_1 controlling for x_2, this estimate is imprecise. Although there are several strategies for dealing with collinear data, none magically extracts nonexistent information from the data. Rather, the research problem is redefined, often implicitly. Sometimes the redefinition is reasonable; usually it is not. The ideal solution to the problem of collinearity is to collect new data in such a manner that the problem is avoided—for example, by experimental manipulation of the xs. Unfortunately, this solution is rarely practical.

Several less adequate strategies for coping with collinear data are described briefly below. I have devoted most space here to variable selection, because selection techniques are commonly abused by social scientists, because the rationale for variable selection is straightforward, and because variable selection is a reasonable approach in certain (limited) circumstances.

Model Respecification. Although collinearity is a data problem, not (necessarily) a deficiency of the model, one approach to the problem is to respecify the model. Perhaps, after further thought, several regressors in the model can be conceptualized as alternative indicators of the same underlying construct. Then the measures can be combined in some manner, or one can be chosen to represent the others. In this context, high correlations among the xs in question indicate high reliability. Imagine an international analysis of factors influencing infant mortality in which GNP (gross national product) per capita, energy use per capita, and televisions per capita are among the independent variables and are highly correlated; a researcher may choose to treat these variables as indicators of the general level of economic development.

Alternatively, we can reconsider whether we really need to control for x_2 (for example) in examining the relationship of y to x_1. Generally, though, respecification of this type is possible only where the original model was poorly thought out or where the researcher is willing to abandon some of the goals of the research. For example, suppose that in a time-series regression examining determinants of married women's labor-force participation, collinearity makes it impossible to separate the effects of men's and women's wage levels. It may still be of interest, however, to determine the partial relationship between women's wage level and labor-force participation controlling for other independent variables in the analysis.

Variable Selection. A common, but usually misguided, approach to collinearity is variable selection, where some procedure is employed to reduce the regressors in the model to a less highly correlated set. Forward stepwise methods add variables to the model one at a time. At each step, the variable that produces the largest increment in R^2 is selected. The procedure stops, for example, when the increment is smaller than a preset criterion. Backward stepwise methods are similar, except that the procedure starts with the full model and deletes variables one at a time. Forward/backward methods combine both approaches.

Stepwise methods frequently are abused by naive researchers who seek to interpret the order of entry of variables into the regression equation as an index of their "importance." That this practice is potentially misleading is suggested by the observation that of two highly correlated independent variables having nearly identical large correlations with y, only one will enter the regression equation, because the other can contribute little additional information. A small modification to the data, or a new sample, could reverse the result.

A technical objection to stepwise methods is that they may fail to turn up the optimal subset of regressors of a given size (i.e., the subset that maximizes R^2). Advances in computer power and in computing procedures make it feasible to examine all subsets of regressors even when k is quite large. Aside from optimizing the selection criterion, subset techniques also have the advantage of revealing alternative, nearly equivalent, models, and thus avoid the misleading appearance of producing a uniquely "correct" result.

One popular approach to subset selection is based on the total (normed) mean-squared error of estimating $E(y)$ from $\hat{y}$—that is, estimating the population regression surface over the observed xs from the fitted surface:

$$\gamma_p = \frac{1}{\sigma^2} \sum_{i=1}^{n} \text{MSE}(\hat{y}_i)$$

$$= \frac{1}{\sigma^2} \sum_{i=1}^{n} \left\{ V(\hat{y}_i) + [E(\hat{y}_i) - E(y_i)]^2 \right\} \qquad [3.4]$$

where the fitted values $\hat{y}_i$ are based on a model containing $p \le k + 1$ regressors (counting the constant, which is always included in the model). Using the error in estimating $E(y)$ as a criterion for model quality is reasonable if the goal is literally to predict y from the xs.

Note that the term $[E(\hat{y}_i) - E(y_i)]^2$ in Equation 3.4 represents the squared bias of $\hat{y}_i$ as an estimator of the population regression surface $E(y_i)$. When collinear regressors are deleted from the model, generally $V(\hat{y}_i)$ will decrease, but—depending on the configuration of data points and the true βs for deleted regressors—bias may be introduced into the fitted values. Because the MSE is the sum of variance and squared bias, the essential question is whether a decrease in variance offsets any increase in bias.

Mallows's (1973) C_p statistic estimates γ_p as

$$C_p = \frac{\Sigma e_i^2}{\hat{\sigma}^2} + 2p - n$$

$$= (k + 1 - p)(F_p - 1) + p$$

where the residuals are from the subset model in question, the error-variance estimate $\hat{\sigma}^2$ is s^2 for the full model, and F_p is the incremental F statistic for testing the hypothesis that the regressors omitted from the current subset have population coefficients of zero. If this hypothesis is true, then $E(F_p) \approx 1$ and thus $E(C_p) \approx p$. A good model, therefore, has C_p close to or below p. As well, minimizing C_p minimizes the sum of squared residuals, and thus maximizes R^2. Note that for the full model, C_{k+1} necessarily equals $k+1$.

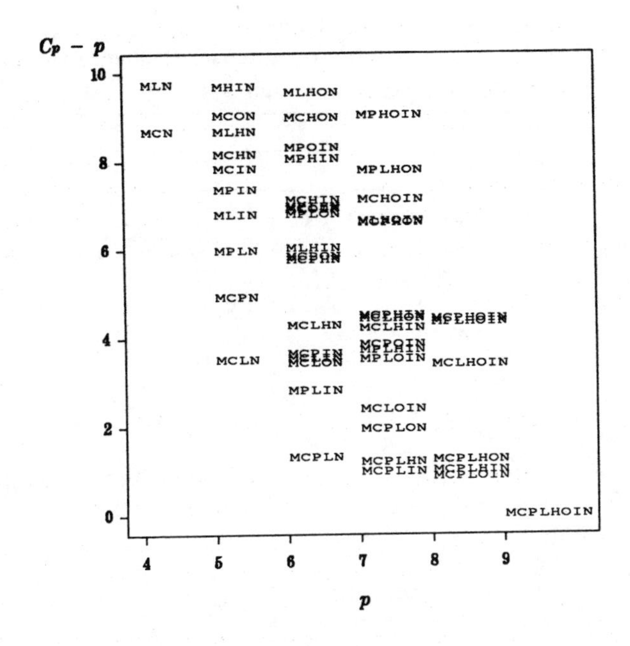

Figure 3.1. Plot of $C_p - p$ versus p for the Census-undercount data. The following capitalized letters are used to label the variables: Minority, Crime, Poverty, Language, High school, hOusing, cIty, and coNventional. Ericksen et al. (1989) selected the independent-variable subset MCN (i.e., Minority, Crime, and Conventional).

Because a good model has C_p close to p, we can identify good models by plotting C_p against p, labeling each point in the plot with a mnemonic representing the independent variables included in the model, and superimposing the line $C_p = p$ on the plot: Good models are close to or below the reference line. The plot is easier to inspect if it is detrended by plotting $C_p - p$ against p (i.e., the reference line is subtracted from each point). Now we can look for models with values of $C_p - p$ near or below zero.

An illustrative detrended C_p plot for the census-undercount data is given in Figure 3.1. Only models for which $C_p - p \leq 10$ (including 52 of the $2^8 - 1 = 255$ predictor subsets) are shown. Ericksen et al. (1989) employed the subset labeled MCN on the plot (with the predictors Minority, Crime, and Conventional). Here, $p = 4$ and $C_p = 12.7$, suggesting

TABLE 3.4
"Best" Subset Regression Models for Ericksen, Kadane, and Tukey's Census-Undercount Data

Predictor	Coefficients[a] $p = 4$	$p = 5$	$p = 6$
Constant	−2.22	−1.98	−0.793
	(0.56)	(0.55)	(0.860)
Minority	0.0786	0.0752	0.101
	(0.0147)	(0.0143)	(0.020)
Crime	0.0363	0.0272	0.0243
	(0.0100)	(0.0104)	(0.0103)
Conventional	0.0280	0.0273	0.0293
	(0.0081)	(0.0077)	(0.0077)
Language		0.209	0.184
		(0.087)	(0.086)
Poverty			−0.110
			(0.062)
R^2	0.638	0.669	0.686
C_p	12.7	8.51	7.32

SOURCE: Data taken from Ericksen, Kadane, and Tukey (1989).

a. Coefficient standard errors are in parentheses.

that there is room for improvement by including more predictors. The regression equation for this subset and the equations for the "best" subsets of four predictors (MCLN, adding Language: $p = 5$ and $C_p = 8.5$) and five predictors (MCPLN, adding Poverty: $p = 6$ and $C_p = 7.3$) appear in Table 3.4. For this data set, backward and forward/backward stepwise procedures identify the "best" subsets of three, four, and five predictors, but the forward method does not. (Recall, however, that Ericksen et al. adopted a more complex estimation strategy than ordinary least-squares regression.)

In applying variable selection it is important to keep the following caveats in mind:

1. Most important, variable selection results in a respecified model that usually does not address the research questions that were posed originally. In particular, if the original model is correctly specified, and if the included and omitted variables are correlated, then coefficient estimates

following variable selection are biased. Consequently, these methods are most useful for pure prediction problems, in which the values of the regressors for the predicted data will be within the configuration for which selection was employed—as in the census-undercount example. In this case, it is possible to get good estimates of $E(y)$ even though the coefficients themselves are biased. If, however, the xs for a new observation are different from those from which the estimates were derived, then the corresponding predicted y might be badly biased.

2. When regressors occur in sets (e.g., of dummy variables), these sets should generally be kept together during selection. Likewise, when there are hierarchical relations among regressors, these relations should be respected: For example, an interaction regressor should not appear in a model that does not contain the main effects marginal to that interaction.
3. Because variable selection optimizes the fit of the model to the sample data, coefficient standard errors calculated following independent-variable selection—and, hence, confidence intervals and hypothesis tests—almost surely overstate the precision of the results. There is, therefore, substantial risk of capitalizing on chance characteristics of the sample. For a solution to this problem, see the discussion of cross-validation in Chapter 10.
4. Variable selection has applications to statistical modeling even when collinearity is not an issue. It is generally unproblematic to eliminate regressors that have small, precisely estimated coefficients, thus producing a more parsimonious model. Indeed, in a very large sample, we may feel justified in deleting regressors with trivially small but "statistically significant" coefficients.

Biased Estimation. Still another general approach to collinear data is biased estimation. The general idea here is to trade a small amount of bias in the coefficient estimates for a substantial reduction in coefficient sampling variance. The result is a smaller mean-squared error of estimation of the βs than provided by the least-squares estimates (cf. the discussion above about estimating $E[y]$ in subset regression). The most common biased estimation method is called *ridge regression* (introduced briefly in Appendix A3.1).

Like variable selection, biased estimation is not a magical panacea for collinearity. For example, ridge regression involves the selection of an arbitrary "ridge constant" controlling the extent to which the ridge estimates differ from least squares: The larger the ridge constant, the greater the bias and the smaller the variance of the ridge estimator. Unfortunately, but reasonably, to pick an optimal ridge constant—or even a good one—generally requires knowledge about the unknown βs that we are trying to estimate. My principal reason for mentioning biased estimation here is to caution *against* its routine use.

Prior Information About the βs. A final approach to estimation with collinear data is to introduce additional prior information that helps to reduce the ambiguity produced by collinearity. There are several different ways that prior information can be brought to bear on a regression, including formal Bayesian analysis, but we shall examine a particularly simple case to illustrate the general point. More complex methods are beyond the scope of this discussion, and are, in any event, difficult to apply in practice (see, e.g., Belsley, Kuh, and Welsch, 1980, pp. 193-204; Theil, 1971, pp. 346-352).

Suppose that we wish to estimate the model

$$y = \beta_0 + \beta_1 x_1 + \beta_2 x_2 + \beta_3 x_3 + \varepsilon$$

where y is savings, x_1 is income from wages and salaries, x_2 is dividend income from stocks, and x_3 is interest income. Imagine that we have trouble estimating β_2 and β_3 because x_2 and x_3 are very highly correlated in our data. Suppose further that we have reason to believe that $\beta_2 = \beta_3$, and denote the common quantity β_*. If x_2 and x_3 were not so highly correlated, then we could reasonably test this belief as a hypothesis. In the current situation, we can fit the model

$$y = \beta_0 + \beta_1 x_1 + \beta_* (x_2 + x_3) + \varepsilon$$

incorporating our belief in the equality of β_2 and β_3 in the specification of the model, and thus eliminating the collinearity problem (along with the possibility of testing the belief).

Comparison of the Approaches. Although I have presented them separately, the several approaches to collinearity have much in common:

1. Model respecification can involve variable selection, and variable selection in effect respecifies the model.
2. Variable selection implicitly constrains the coefficients of deleted regressors to zero.
3. Variable selection produces biased coefficient estimates if the deleted variables have nonzero βs and are correlated with the included variables.

4. Certain types of prior information (as in the hypothetical example) result in a respecified model.
5. It can be demonstrated that biased-estimation methods like ridge regression implicitly place prior constraints on the values of the βs.

The primary lesson to be drawn from these comparisons is that mechanical model-selection and modification procedures disguise the substantive implications of modeling decisions. Consequently, these methods generally cannot compensate for weaknesses in the data and are no substitute for judgment and thought.

4. OUTLYING AND INFLUENTIAL DATA

Unusual data are problematic in a least-squares regression because they can unduly influence the results of the analysis, and because their presence may be a signal that the regression model fails to capture important characteristics of the data. Some central distinctions are illustrated in Figure 4.1 for the simple-regression model $y = \beta_0 + \beta_1 x + \varepsilon$.

In simple regression, an outlier is an observation whose dependent-variable value is unusual given the value of the independent variable. In contrast, a univariate outlier is a value of y or x that is unconditionally unusual; such a value may or may not be a regression outlier. Regression outliers appear in both part a and part b of Figure 4.1. In Figure 4.1a, the outlying observation has an x value at the center of the x distribution; as a consequence, deleting the outlier has no impact on the least-squares slope b_1 and little impact on the intercept b_0. In Figure 4.1b, the outlier has an unusual x value, and consequently its deletion markedly affects both the slope and the intercept. Because of its unusual x value, the last observation in Figure 4.1b has strong leverage on the regression coefficients, whereas the middle observation in Figure 4.1a is at a low-leverage point.

The combination of high leverage with an outlier produces substantial influence on the regression coefficients. In Figure 4.1c, the last observation has no influence on the regression coefficients even though it is a high-leverage point, because this observation is not out of line with the rest of the data. The following heuristic formula helps to distinguish among these concepts:

$$\text{Influence on Coefficients} = \text{Leverage} \times \text{Discrepancy}$$

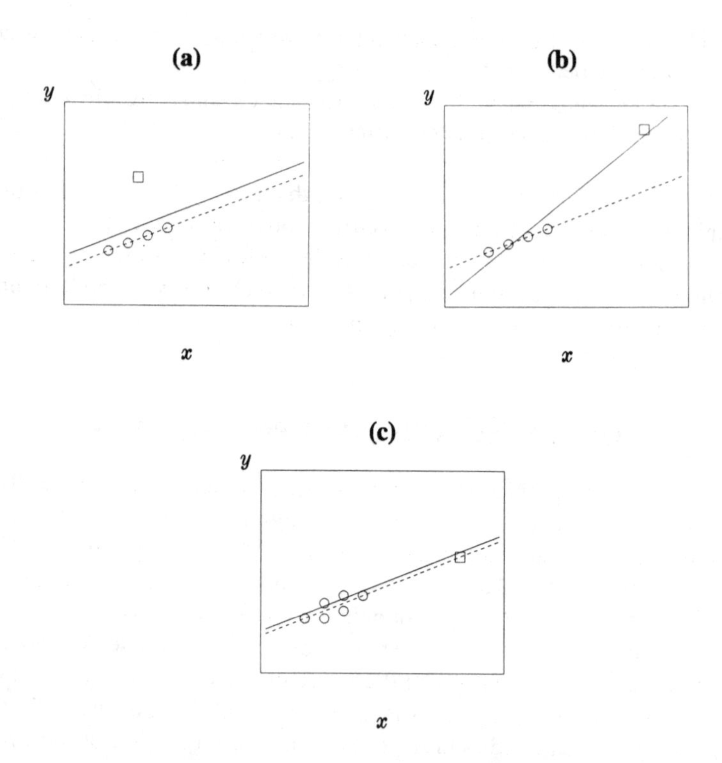

Figure 4.1. Leverage and influence in simple-regression analysis. (a) An outlier near the mean of x has little influence on the regression coefficients. (b) An outlier far from the mean of x markedly affects the regression coefficients. (c) A high-leverage observation in line with the rest of the data does not influence the regression coefficients.

A simple and transparent example, with real data from Davis (1990), appears in Figure 4.2. These data record the measured and reported weight (in kilograms) of 183 male and female subjects who engage in programs of regular physical exercise. As part of a larger study, the investigator was interested in ascertaining whether the subjects reported their weights accurately, and whether men and women reported similarly. (The published study is based on the data for the female subjects only and includes additional data for non-exercising women.) Davis (1990) gives the correlation between measured and reported weight.

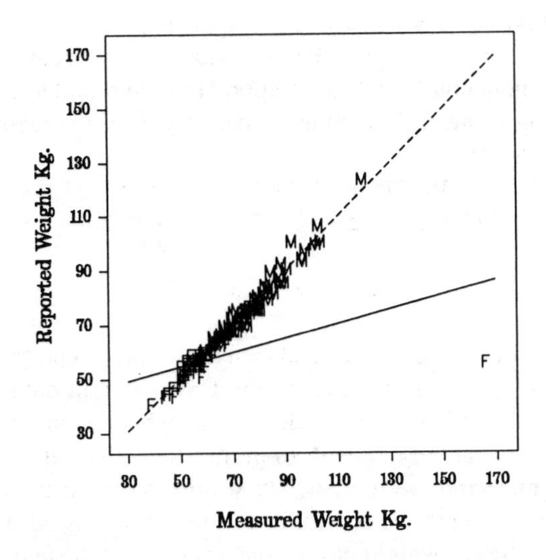

Figure 4.2. Regression of reported weight in kilograms on measured weight and gender for 183 subjects engaged in regular exercise. The solid line shows the least-squares regression for women, the broken line the regression for men.
SOURCE: Data taken from C. Davis, personal communication.

A least-squares regression of reported weight (*RW*) on measured weight (*MW*), a dummy variable for sex (*F*: coded one for women, zero for men), and an interaction regressor produces the following results (with coefficient standard errors in parentheses):

$$\hat{RW} = \underset{(3.28)}{1.36} + \underset{(0.043)}{0.990}MW + \underset{(3.9)}{40.0}F - \underset{(0.056)}{0.725}MW \times F$$

$$R^2 = 0.89 \qquad s = 4.66$$

Were these results to be taken seriously, we would conclude that men are on average accurate reporters of their weights (because $b_0 \approx 0$ and $b_1 \approx 1$), whereas women tend to overreport their weights if they are relatively light and underreport if they are relatively heavy. However, Figure 4.2 makes clear that the differential results for women and men are due to one female subject whose reported weight is about average (for women), but whose measured weight is extremely large.

In fact, this subject's measured weight and height (in centimeters) were switched erroneously on data entry, as Davis discovered after calculating an anomalously low correlation between reported and measured weight among women. Correcting the data produces the regression

$$\hat{RW} = \underset{(1.58)}{1.36} + \underset{(0.021)}{0.990\,MW} + \underset{(2.45)}{1.98\,F} - \underset{(0.0385)}{0.0567\,MW \times F}$$

$$R^2 = 0.97 \qquad s = 2.24$$

which suggests that both women and men are accurate reporters of weight.

There is another way to analyze the Davis weight data: One of the investigator's interests was to determine whether subjects reported their weights accurately enough to permit the substitution of reported weight for measured weight, which would decrease the cost of collecting data on weight. It is natural to think of reported weight as influenced by "real" weight, as in the regression presented above in which reported weight is the dependent variable. The question of substitution, however, is answered by the regression of measured weight on reported weight, giving the following results for the *uncorrected* data:

$$\hat{MW} = \underset{(5.92)}{1.79} + \underset{(0.076)}{0.969\,RW} + \underset{(9.30)}{2.07\,F} - \underset{(0.147)}{0.00953\,RW \times F}$$

$$R^2 = 0.70 \qquad s = 8.45$$

Note that here the outlier does not have much impact on the regression coefficients, precisely because the value of RW for this observation is near $\overline{RW}$ for women. However, there is a marked effect on the multiple correlation and standard error: For the corrected data, $R^2 = 0.97$, $s = 2.25$.

Measuring Leverage: Hat-Values

The so-called *hat-value* h_i is a common measure of leverage in regression. These values are so named because it is possible to express the fitted values $\hat{y}_j$ in terms of the observed values y_i:

$$\hat{y}_j = h_{1j}y_1 + h_{2j}y_2 + \ldots + h_{jj}y_j + \ldots + h_{nj}y_n = \sum_{i=1}^{n} h_{ij}y_i$$

Thus the weight h_{ij} captures the extent to which y_i can affect $\hat{y}_j$: If h_{ij} is large, then the ith observation can have a substantial impact on the jth fitted value. It may be shown that $h_{ii} = \Sigma_{j=1}^{n} h_{ij}^2$ and so the hat-value $h_i = h_{ii}$ summarizes the potential influence (the leverage) of y_i on all of the fitted values. The hat-values are bounded between $1/n$ and 1 (i.e., $1/n \le h_i \le 1$), and the average hat-value is $\bar{h} = (k + 1)/n$ (see Appendix A4.1).

In simple-regression analysis, the hat-values measure distance from the mean of x:

$$h_i = \frac{1}{n} + \frac{(x_i - \bar{x})^2}{\sum_{j=1}^{n} (x_j - \bar{x})^2}$$

In multiple regression, h_i measures distance from the centroid (point of means) of the xs, taking into account the correlational structure of the xs, as illustrated for $k = 2$ in Figure 4.3. Multivariate outliers in the x space are thus high-leverage observations.

For Davis's regression of reported weight on measured weight, the largest hat-value by far belongs to the 12th subject, whose measured weight was erroneously recorded as 166 kg: $h_{12} = 0.714$. This quantity is many times the average hat-value, $\bar{h} = (3 + 1)/183 = 0.0219$.

Detecting Outliers: Studentized Residuals

To identify an outlying observation, we need an index of the unusualness of y given the xs. Generally, discrepant observations have large residuals, but it turns out that even if the errors ε_i have equal variances (as assumed in the regression model), the residuals e_i do not: $V(e_i) = \sigma^2(1 - h_i)$ (see Appendix A4.2). High-leverage observations, therefore, tend to have small residuals—a sensible result, because these observations can force the regression surface to be close to them.

Although we can form a standardized residual by calculating $e_i' = e_i / s\sqrt{1 - h_i}$, this measure suffers from the defect that the numerator and denominator are not independent, preventing e_i' from following a t distribution: When $|e_i|$ is large, $s = \sqrt{\Sigma e_i^2/(n - k - 1)}$, which contains e_i^2, tends to be large as well. Suppose, however, that we refit the regression model deleting the ith observation, obtaining an estimate $s_{(-i)}$ of σ based on the rest of the data. Then the *studentized residual*

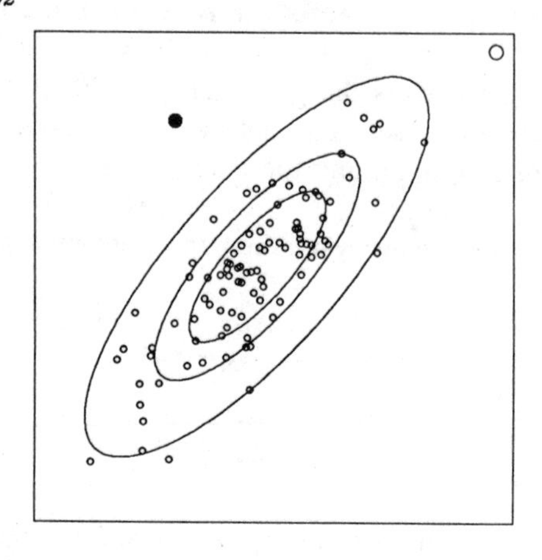

Figure 4.3. Contours of constant leverage (constant h_i) for $k = 2$ independent variables. Two high-leverage points appear: One (shown as a large hollow dot) has unusually large values for each of x_1 and x_2, but the other (large filled dot) is unusual only in its combination of x_1 and x_2 values.

$$t_i = \frac{e_i}{s_{(-i)} \sqrt{1 - h_i}} \qquad [4.1]$$

has independent numerator and denominator, and follows a t distribution with $n - k - 2$ degrees of freedom.

An alternative, but equivalent, procedure for finding the studentized residuals employs the "mean-shift" outlier model

$$y_j = \beta_0 + \beta_1 x_{1j} + \ldots + \beta_k x_{kj} + \gamma d_j + \varepsilon_j \qquad [4.2]$$

where d is a dummy variable set to one for observation i and zero for all other observations. Thus

$$E(y_j) = \beta_0 + \beta_1 x_{1j} + \ldots + \beta_k x_{kj} \quad \text{for } j \neq i$$

$$E(y_i) = \beta_0 + \beta_1 x_{1i} + \ldots + \beta_k x_{ki} + \gamma$$

It would be natural to specify Equation 4.2 if before examining the data we suspected that observation i differed from the others. Then, to test H_0: $\gamma = 0$, we would find $t_i = \hat{\gamma} / \mathrm{SE}(\hat{\gamma})$, which is distributed as t_{n-k-2} under H_0, and which (it turns out) is the studentized residual of Equation 4.1.

Here, as elsewhere in statistics, terminology is not wholly standard: t_i is sometimes called a *deleted studentized residual*, an *externally studentized residual*, or even a *standardized residual*. Because the last term also is often applied to e_i', it is important to determine exactly what is being calculated by a computer program before using these quantities. In large samples, though, usually $t_i \approx e_i' \approx e_i / s$.

Testing for Outliers in Regression. Because in most applications we do not suspect a particular observation in advance, we can in effect refit the mean-shift model n times, once for each observation, producing $t_1, t_2, \ldots, t_n$. In practice, alternative formulas to Equations 4.1 and 4.2 provide the t_i with little computational effort. Usually, our interest then will focus on the largest absolute t_i, called t^*. Because we have picked the biggest of n test statistics, however, it is no longer legitimate simply to use t_{n-k-2} to find the statistical significance of t^*: For example, even if our model is wholly adequate, and disregarding for the moment the dependence among the t_is, we would expect to observe about 5% of t_is beyond $t_{0.025} \approx \pm 2$, about 1% beyond $t_{0.005} \approx \pm 2.6$, and so forth.

One solution to the problem of simultaneous inference is to perform a Bonferroni adjustment to the p value for the largest t_i. (Another way to take into account the number of studentized residuals, by constructing a quantile-comparison plot, is discussed in Chapter 5.) The Bonferroni test requires either a special t table or, more conveniently, a computer program that returns accurate p values for t far into the tail of the distribution. In the latter event, suppose that $p' = \Pr(t_{n-k-2} > |t^*|)$. Then the p value for testing the statistical significance of t^* is $p = 2np'$. The factor 2 reflects the two-tail character of the test: We want to detect large negative as well as large positive outliers. The factor n adjusts for conducting n simultaneous tests, which is implicit in selecting the largest of n test statistics. Beckman and Cook (1983) have shown that the Bonferroni adjustment usually is exact for testing the largest studentized residual. Note that a much larger t^* is required for a statistically significant result than would be the case for an ordinary individual t test.

In Davis's regression of reported weight on measured weight, the largest studentized residual by far belongs to the 12th observation: $t_{12} = -24.3$. Here, $n - k - 2 = 183 - 3 - 2 = 178$, and $\Pr(t_{178} > 24.3) << 10^{-8}$. (The symbol “<<” means “much less than.” The computer program that I employed to find the tail probability was unable to calculate a more accurate result for such a large t.) The Bonferroni p value for the outlier test is $p << 178 \times 2 \times 10^{-8} = 4 \times 10^{-6}$ (i.e., 0.000004), an unambiguous result.

An Analogy to Insurance. Thus far, I have treated the identification (and, implicitly, the potential correction, removal, or accommodation) of outliers as a hypothesis-testing problem. Although this is by far the most common approach in practice, a more reasonable general perspective weighs the costs and benefits for estimation of rejecting a potentially outlying observation.

Suppose, for the moment, that the observation with the largest t_i is simply an unusual data point, but one generated by the assumed statistical model, that is, $y_i = \beta_0 + \beta_1 x_{1i} + \ldots + \beta_k x_{ki} + \varepsilon_i$, with $\varepsilon_i \sim \text{NID}(0, \sigma^2)$. To discard an observation under these circumstances would decrease the efficiency of estimation, because when the model—including the assumption of normality—is correct, the least-squares estimator is maximally efficient among all unbiased estimators of the βs. If, however, the datapoint in question does not belong with the rest (say, e.g., the mean-shift model applies), then to eliminate it may make estimation more efficient. Anscombe (1960) expressed this insight by drawing an analogy to insurance: To obtain protection against “bad” data, one purchases a policy of outlier rejection (or uses an estimator that is resistant to outliers—a so-called *robust* estimator), a policy paid for by a small premium in efficiency when the policy rejects “good” data.

Let P denote the desired premium, say 0.05—a 5% increase in estimator mean-squared error if the model holds for all of the data. Let z represent the unit-normal deviate corresponding to a tail probability of $P(n-k-1)/n$. Following the procedure derived by Anscombe and Tukey (1963), compute $m = 1.4 + 0.85z$, and then find

$$f = m\left(1 - \frac{m^2 - 2}{4(n - k - 1)}\right) \times \sqrt{\frac{n - k - 1}{n}} \qquad [4.3]$$

and

$$t' = \frac{f\sqrt{n-k-2}}{\sqrt{n-k-1-f^2}} \quad [4.4]$$

Finally, reject the observation with the largest studentized residual if $|t^*| > t'$. In a real application, of course, we should inquire about discrepant observations (see the discussion at the end of this section).

For example, for Davis's first regression $n = 183$ and $k = 3$; so for a premium of $P = 0.05$, we have

$$P(n-k-1)/n = 0.05\,(183-3-1)/183 = 0.0489$$

From the unit-normal table, $z = 1.66$, from which $m = 1.4 + 0.85 \times 1.66 = 2.81$. Then, using Equation 4.3, $f = 2.76$, and using Equation 4.4, $t' = 2.81$. Because $t^* = 24.3$ is much larger than t', the 12th observation is identified as an outlier.

Measuring Influence: Cook's Distance and Other Diagnostics

As noted previously, influence on the regression coefficients combines leverage and discrepancy. The most direct measure of influence simply examines the impact on each coefficient of deleting each observation in turn:

$$d_{ij} = b_j - b_{j(-i)}, \quad \text{for } i = 1, \ldots, n; \ \ j = 0, \ldots, k$$

where $b_{j(-i)}$ denotes the least-squares estimate of β_j produced when the ith observation is omitted. To assist in interpretation, it is useful to scale the d_{ij} by (deleted) estimates of the coefficient standard errors:

$$d^*_{ij} = \frac{d_{ij}}{\text{SE}_{(-i)}(b_j)}$$

Following Belsley et al. (1980), the d_{ij} are often termed DFBETA_{ij}, and the d^*_{ij} are called DFBETAS_{ij}.

One problem associated with using the d_{ij} or d^*_{ij} is their large number: $n(k + 1)$ of each. Of course, these values can be more quickly examined graphically than in numerical tables. For example, we can construct an "index plot" of the d^*_{ij}s for each coefficient $j = 0, 1, \ldots, k$—simple scatterplots with d^*_{ij} on the vertical axis versus the observation index i on the horizontal axis. Nevertheless, it is useful to have a summary index of the influence of each observation on the fit.

Cook (1977) has proposed measuring the "distance" between the b_j and the corresponding $b_{j(-i)}$ by calculating the F statistic for the "hypothesis" that $\beta_j = b_{j(-i)}, j = 0, 1, \ldots, k$. This statistic is recalculated for each observation $i = 1, \ldots, n$. The resulting values should not literally be interpreted as F tests—Cook's approach merely exploits an analogy to testing to produce a measure of distance independent of the scales of the x variables. Cook's statistic may be written (and simply calculated) as

$$D_i = \frac{e_i'^2}{k+1} \times \frac{h_i}{1-h_i}$$

In effect, the first term is a measure of discrepancy, and the second a measure of leverage (see Appendix A4.3). We look for values of D_i that are substantially larger than the rest.

Belsley et al. (1980) have suggested the very similar measure

$$\text{DFFITS}_i = t_i \sqrt{\frac{h_i}{1-h_i}}$$

Note that except for unusual data configurations $D_i \approx \text{DFFITS}_i^2/(k+1)$. Other global measures of influence are available (see Chatterjee and Hadi, 1988, Ch. 4, for a comparative treatment).

Because all of the deletion statistics depend on the hat-values and residuals, a graphical alternative to either of the general influence measures is to plot the h_i against the t_i and to look for observations for which both are big. A slightly more sophisticated version of this plot displays circles of area proportional to Cook's D instead of points (see Figure 4.6, page 38). We can follow up by examining the d_{ij} or d_{ij}^* for the observations with the largest few D_i's, $|\text{DFFITS}_i|$, or combination of large h_i and $|t_i|$.

For Davis's regression of reported weight on measured weight, all of the indices of influence point to the obviously discrepant 12th observation:

$$\text{Cook's } D_{12} = 85.9 \text{ (next largest, } D_{21} = 0.065)$$

$$\text{DFFITS}_{12} = -38.4 \text{ (next largest, DFFITS}_{50} = 0.512)$$

$$\text{DFBETAS}_{0,12} = \text{DFBETAS}_{1,12} = 0, \text{ DFBETAS}_{2,12} = 20.0,$$
$$\text{DFBETAS}_{3,12} = -24.8$$

Note that observation 12, which is for a female subject, has no impact on the male intercept b_0 and slope b_1.

Influence on Standard Errors. In developing the concept of influence in regression, I have focused on changes in regression coefficients. Other regression outputs may be examined as well, however. One important output is the set of coefficient variances and covariances, which capture the precision of estimation. For example, recall Figure 4.1c, where a high-leverage observation exerts no influence on the regression coefficients, because it is in line with the rest of the data. The estimated standard error of the least-squares slope in simple regression is $\text{SE}(b_1) = s/\sqrt{\Sigma (x_i - \bar{x})^2}$, and, therefore, by increasing the variance of x the high-leverage observation serves to decrease $\text{SE}(b_1)$, even though it does not influence b_0 and b_1. Depending on context, such an observation may be considered beneficial—increasing the precision of estimation—or it may cause us to exaggerate our confidence in the estimate b_1.

In multiple regression, we can examine the impact of deleting each observation in turn on the size of the joint-confidence region for the βs. Recall from Chapter 2 that the size of this region is analogous to the length of a confidence interval for an individual coefficient, which in turn is proportional to coefficient standard error. The squared length of a confidence interval is therefore proportional to coefficient sampling variance, and, analogously, the squared size of a joint confidence region is proportional to the "generalized" variance of a set of coefficients. An influence measure proposed by Belsley et al. (1980) closely approximates the squared ratio of volumes of the deleted and full-data confidence regions:

$$\text{COVRATIO}_i = \frac{1}{(1 - h_i)\left(\dfrac{n - k - 2 + t_i^2}{n - k - 1}\right)^{k+1}}$$

Alternative, similar measures have been suggested by several authors (again, Chatterjee and Hadi, 1988, Ch. 4, provide a comparative discussion). Look for values of COVRATIO_i that differ substantially from 1.

As for measures of influence on the regression coefficients, both the hat-value and the (studentized) residual figure in COVRATIO. A large hat-value produces a large COVRATIO, however, even when (actually, especially when) t is small, because a high-leverage, in-line observation improves the precision of estimation. In contrast, a discrepant, low-leverage observation might not change the coefficients much, but it decreases the precision of estimation by increasing the estimated error variance; such an observation, with small h and large t, produces a COVRATIO substantially below 1.

For example, for Davis's first regression by far the most extreme value is $\text{COVRATIO}_{12} = 0.0103$. In this case, a very large $h_{12} = 0.714$ is more than offset by a massive $t_{12} = -24.3$.

Influence on Collinearity. Other characteristics of a regression analysis also may be influenced by individual observations, including the degree of collinearity. Although a formal consideration of influence on collinearity is above the level of this presentation (see Chatterjee and Hadi, 1988, Ch.4-5), the following remarks may prove helpful:

1. Influence on collinearity is one of the factors reflected in influence on coefficient standard errors. Influence on the error variance and influence on the variation of the xs also are implicitly factored into a measure such as COVRATIO, however. As well, COVRATIO and similar measures examine the sampling variances and covariances of all of the regression coefficients, including the constant. Nevertheless, our concern for collinearity reflects its impact on the precision of estimation, and the global precision of estimation is assessed by COVRATIO.
2. Collinearity-influential points are those that either induce or substantially weaken correlations among the xs. Such points usually—but not always—have large hat-values. Conversely, points with large hat-values often influence collinearity.
3. Individual points that induce collinearity are obviously problematic. Points that substantially weaken collinearity also merit examination, because they may cause us to be overly confident in our results.
4. It is frequently possible to detect collinearity-influential observations by plotting independent variables against each other. This approach will fail, however, if the collinear relations in question involve more than two independent variables at a time.

Numerical Cutoffs for Diagnostic Statistics

I have deliberately refrained from suggesting specific numerical criteria for identifying noteworthy observations on the basis of measures of leverage and influence. I believe that generally it is more useful to examine the distributions of these quantities to locate observations with unusual values. For studentized residuals, the hypothesis-testing and insurance approaches produce cutoffs of sorts, but even these numerical criteria are no substitute for graphical examination of the residuals.

Nevertheless, cutoffs can be of some use, as long as they are not given too much weight, and especially when they serve to enhance graphical displays. A horizontal line may be drawn on an index plot, for example, to draw attention to values beyond a cutoff. Similarly, such values may be identified individually in a graph (as in Figure 4.6, page 38).

Cutoffs for a diagnostic statistic may be the product of statistical theory, or they may result from examination of the sample distribution of the statistic. Cutoffs may be absolute, or they may be adjusted for sample size (Belsley et al., 1980, Ch. 2). For some diagnostic statistics, such as measures of influence, absolute cutoffs are unlikely to identify noteworthy observations in large samples. In part, this characteristic reflects the ability of large samples to absorb discrepant data without changing the results substantially, but it is still often of interest to identify *relatively* influential points, even if no observation has strong *absolute* influence.

The cutoffs presented below are, as explained briefly, based on the application of statistical theory. An alternative, very simple, and universally applicable data-based criterion is to examine the most extreme 5% (say) of values for a diagnostic measure.

1. *Hat-values:* Belsley et al. (1980) suggest that hat-values exceeding about twice the average $(k + 1)/n$ are noteworthy. This size-adjusted cutoff was derived as an approximation identifying the most extreme 5% of cases when the xs are multivariate-normal and k and $n - k - 1$ are relatively large, but it is recommended by these authors as a rough general guide. (See Chatterjee and Hadi [1988, Ch. 4] for a discussion of alternative cutoffs for hat-values.)
2. *Studentized residuals:* Beyond the issues of "statistical significance" and estimator robustness and efficiency discussed above, it sometimes helps to call attention to residuals that are relatively large. Recall that under ideal conditions about 5% of studentized residuals are outside the range $|t_i| \leq 2$. It is therefore reasonable, for example, to draw lines at ± 2 on a display of studentized residuals to highlight observations outside this range.

3. *Measures of influence:* Many cutoffs have been suggested for different measures of influence. A few are presented here:
   a. *Standardized change in regression coefficients:* The d_{ij}^* are scaled by standard errors, and consequently $|d_{ij}^*| > 1$ or 2 suggests itself as an absolute cutoff. As explained above, however, this criterion is unlikely to nominate observations in large samples. Belsley et al. propose the size-adjusted cutoff $2/\sqrt{n}$ for identifying noteworthy d_{ij}^*s.
   b. *Cook's* D *and DFFITS:* A variety of numerical cutoffs have been recommended for Cook's D and DFFITS—exploiting the analogy between D and an F statistic, for example. Chatterjee and Hadi (1988) suggest comparing $|\text{DFFITS}_i|$ with the size-adjusted cutoff $2\sqrt{(k+1)/(n-k-1)}$. (Also see Cook [1977], Belsley et al. [1980], and Velleman and Welsch [1981].) Moreover, because of the approximate relationship between DFFITS and Cook's D, it is simple to translate cutoffs between the two measures. For Chatterjee and Hadi's criterion, for example, we have the translated cutoff $D_i > 4/(n-k-1)$. Absolute cutoffs, such as $D_i > 1$, risk missing influential data.
   c. *COVRATIO:* Belsley et al. suggest that COVRATIO_i is noteworthy when $|\text{COVRATIO}_i - 1|$ exceeds the size-adjusted cutoff $3(k+1)/n$.

Jointly Influential Subsets of Observations: Partial-Regression Plots

As illustrated in Figure 4.4, subsets of observations can be jointly influential or can offset each other's influence. Often, influential subsets or multiple outliers can be identified by applying single-observation diagnostics sequentially. It is important, however, to refit the model after deleting each point, because the presence of a single influential value may dramatically affect the fit at other points. Still, the sequential approach is not always successful.

Although it is possible to generalize deletion statistics formally to subsets of several points, the very large number of subsets (there are $n!/[p!(n-p)!]$ subsets of size p) usually renders the approach impractical (but see Belsley et al. 1980, Ch. 2; Chatterjee and Hadi, 1988, Ch. 5). An attractive alternative is to employ graphical methods.

A particularly useful influence graphic is the partial-regression plot, also called a *partial-regression leverage plot* or an *added-variable plot*. Let $y_i^{(1)}$ represent the residuals from the least-squares regression of y on all of the xs save x_1, that is, the residuals from the fitted model

$$y_i = b_0^{(1)} + b_2^{(1)} x_{2i} + \ldots + b_k^{(1)} x_{ki} + y_i^{(1)}$$

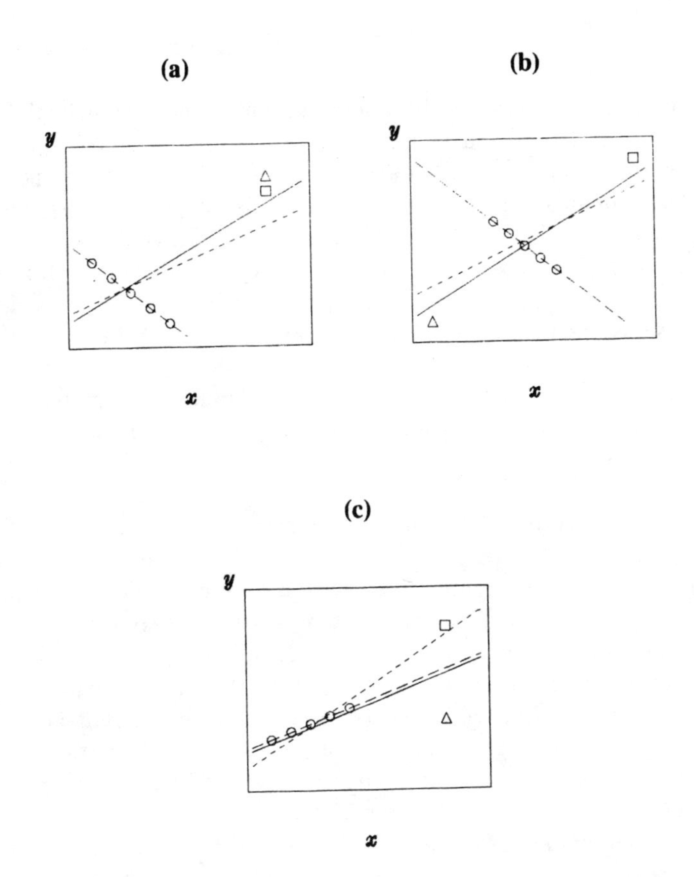

Figure 4.4. Jointly influential data. In each case, the solid line gives the regression for all of the data, the light broken line gives the regression with the triangle deleted, and the heavy broken line gives the regresson with both the square and the triangle deleted. (a) Jointly influential observations located close to one another: Deletion of both observations has a much greater impact than deletion of only one. (b) Jointly influential observations located on opposite sides of the data. (c) Observations that offset one another: The regression with both observations deleted is the same as for the whole dataset.

Likewise, the $x_i^{(1)}$ are residuals from the least-squares regression of x_1 on the other xs:

$$x_{1i} = c_0^{(1)} + c_2^{(1)} x_{2i} + \ldots + c_k^{(1)} x_{ki} + x_i^{(1)}$$

The notation emphasizes the interpretation of the residuals $y^{(1)}$ and $x^{(1)}$ as the parts of y and x_1 that remain when the effects of $x_2, \ldots, x_k$ are removed. It may be shown (see Appendix A4.4) that the slope from the least-squares regression of $y^{(1)}$ on $x^{(1)}$ is simply the least-squares slope b_1 from the full multiple regression, and that the residuals from this regression are the same as those from the full regression, that is, $y_i^{(1)} = b_1 x_i^{(1)} + e_i$. Note that no constant is required here, because as least-squares residuals, both $y^{(1)}$ and $x^{(1)}$ have means of zero.

Plotting $y^{(1)}$ against $x^{(1)}$ permits us to examine leverage and influence on b_1. Similar partial-regression plots can be constructed for the other regression coefficients, including b_0:

$$\text{Plot } y^{(j)} \text{ versus } x^{(j)}, \text{ for } j = 0, 1, \ldots, k$$

In the case of b_0, we regress the "constant regressor" $x_0 = 1$ and y on x_1 through x_k, with no constant in the regression equations.

Illustrative partial-regression plots appear in Figure 4.5. The data for this example are drawn from Duncan (1961), who regressed the rated prestige of 45 occupations (P, assessed as the percentage of raters scoring the occupations as "good" or "excellent") on the income and educational levels of the occupations in 1950 (respectively, I, the percent of males earning at least \$3,500, and E, the percent of male high-school graduates). The primary aim of this regression was to produce fitted prestige scores for occupations for which there were no direct prestige ratings, but for which income and educational data were available. The fitted regression (with standard errors in parentheses) is

$$\begin{aligned} \hat{P} = &-6.06 + 0.599I + 0.546E \\ &(4.27) \quad (0.120) \quad (0.098) \end{aligned}$$

$$R^2 = 0.83 \qquad s = 13.4$$

The partial-regression plot for income (Figure 4.5a) reveals three apparently influential observations that serve to decrease the income slope: ministers (6), whose income is unusually low given the educational level of the occupation; and railroad conductors (16) and rail-

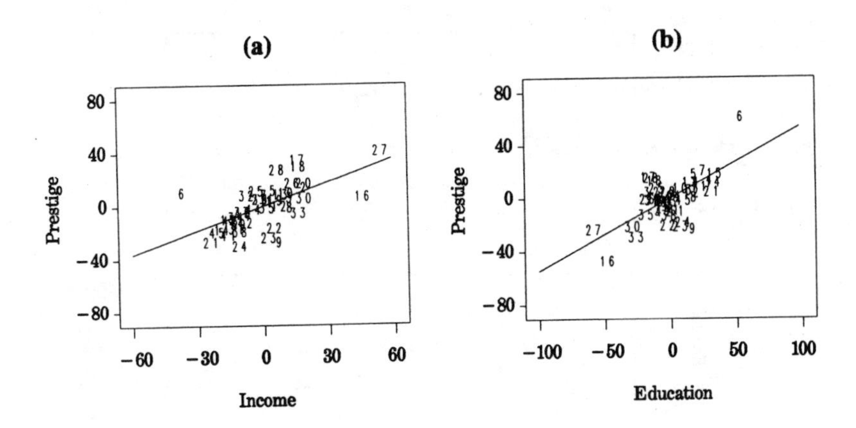

Figure 4.5. Partial-regression plots for (a) income and (b) education in the regression of prestige on the income and education levels of 45 U.S. occupations in 1950. The observation numbers of the points are plotted. If the plots were drawn to a larger scale, as on a computer screen, then the names of the occupations could be plotted in place of their numbers. The partial-regression plot for the constant is not shown.

road engineers (27), whose incomes are unusually high given education. Recall that the horizontal variable in the partial-regression plot is the residual from the regression of income on education, and thus values far from zero in this direction are those for which income is unusual given education.

The partial-regression plot for education (Figure 4.5b) shows that the same three observations have relatively high leverage on the education coefficient: Observations 6 and 16 tend to increase b_2, whereas observation 27 appears to be closer in line with the rest of the data.

Examining the single-observation deletion diagnostics reveals that observation 6 has the largest Cook's D ($D_6 = 0.566$) and studentized residual ($t_6 = 3.14$). This studentized residual is not especially big, however: The Bonferroni p value for the outlier test is $\Pr(t_{41} > 3.14) \times 2 \times 45 = 0.14$. Figure 4.6 displays a plot of studentized residuals versus hat-values, with the areas of the plotted circles proportional to values of Cook's D. Observation indices are shown on the plot for $|t_i| > 2$ or $h_i > 2(k + 1)/n = 2(2 + 1)/45 = 0.13$.

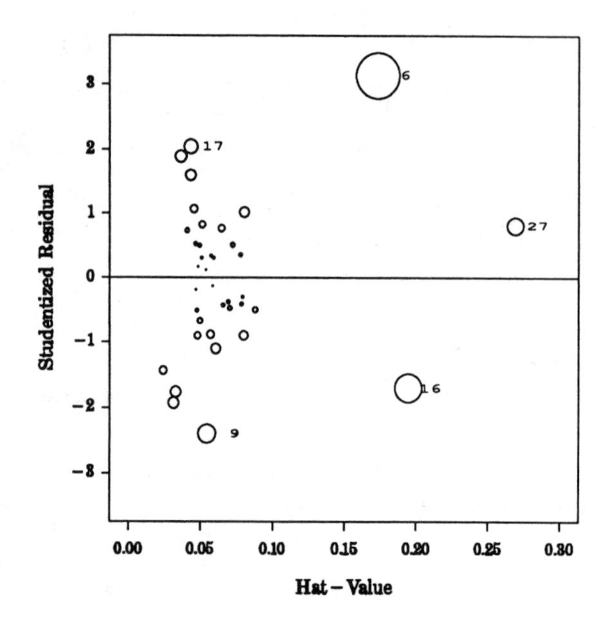

Figure 4.6. Plot of studentized residuals against hat-values for the regression of occupational prestige on income and education. Each point is plotted as a circle with area proportional to Cook's D. The observation number is shown when $h_i > 2\bar{h} = 0.13$ or $|t_i| > 2$.

Deleting observations 6 and 16 produces the fitted regression

$$\hat{P} = -6.41 + 0.867I + 0.332E$$
$$(3.65) \quad (0.122) \quad (0.099)$$

$$R^2 = 0.88 \qquad s = 11.4$$

which, as expected from the partial-regression plots, has a larger income slope and smaller education slope than the original regression. The estimated standard errors are likely optimistic, because relative outliers have been trimmed away. Deleting observation 27 as well further increases the income slope and decreases the education slope, but the change is not dramatic: $b_I = 0.931$, $b_E = 0.285$.

Should Unusual Data Be Discarded?

The discussion in this section has proceeded as if outlying and influential data are simply discarded. Though problematic data should not be ignored, they also should not be deleted automatically and thoughtlessly:

1. It is important to investigate why data are unusual. Truly bad data (e.g., errors in data entry as in Davis's regression) can often be corrected or, if correction is not possible, thrown away. Alternatively, when a discrepant data-point is correct, we may be able to understand why the observation is unusual. For Duncan's regression, for example, it makes sense that ministers enjoy prestige not accounted for by the income and educational levels of the occupation. Likewise, I suspect that the high incomes of railroad workers relative to their educational level and prestige reflect the power of railroad unions around 1950. In a case like this, we may choose to deal with outlying observations separately.

2. Alternatively, outliers or influential data may motivate model respecification. For example, the pattern of outlying data may suggest introduction of additional independent variables. If, in Duncan's regression, we can identify a factor that produces the unusually high prestige of ministers (net of their income and education), and we can measure that factor for other occupations, then this variable could be added to the regression. In some instances, transformation of the dependent variable or of an independent variable may, by rendering the error distribution symmetric or eliminating nonlinearity (see Chapters 5 and 7), draw apparent outliers toward the rest of the data. We must, however, be careful to avoid "overfitting" the data—permitting a small portion of the data to determine the form of the model. I shall return to this problem in Chapters 9 and 10.

3. Except in clear-cut cases, we are justifiably reluctant to delete observations or to respecify to accommodate unusual data. Some researchers reasonably adopt alternative estimation strategies, such as robust regression, which continuously downweights outlying data rather than simply including or discarding them. Such methods are termed "robust" because they behave well even when the errors are not normally distributed (see the discussion of lowess in Appendix A6.1 for an example). As mentioned in passing, the attraction of

robust estimation may be understood using Anscombe's insurance analogy: Robust methods are nearly as efficient as least squares when the errors are normally distributed, and much more efficient in the presence of outliers. Because these methods assign zero or very small weight to highly discrepant data, however, the result is not generally very different from careful application of least squares, and, indeed, robust-regression weights may be used to identify outliers. Moreover, most robust-regression methods are vulnerable to high-leverage points (but see the "high-breakdown" estimators described by Rousseeuw and Leroy, 1987).

5. NON-NORMALLY DISTRIBUTED ERRORS

The assumption of normally distributed errors is almost always arbitrary. Nevertheless, the central-limit theorem assures that under very broad conditions inference based on the least-squares estimators is approximately valid in all but small samples. Why, then, should we be concerned about non-normal errors?

First, although the validity of least-squares estimation is robust—as stated, the levels of tests and confidence intervals are approximately correct in large samples even when the assumption of normality is violated—the method is not robust in efficiency: The least-squares estimator is maximally efficient among unbiased estimators when the errors are normal. For some types of error distributions, however, particularly those with heavy tails, the efficiency of least-squares estimation decreases markedly. In these cases, the least-squares estimator becomes much less efficient than alternatives (e.g., so-called robust estimators, or least-squares augmented by diagnostics). To a substantial extent, heavy-tailed error distributions are problematic because they give rise to outliers, a problem that I addressed in the previous chapter.

A commonly quoted justification of least-squares estimation—called the *Gauss-Markov theorem*—states that the least-squares coefficients are the most efficient unbiased estimators that are *linear* functions of the observations y_i. This result depends on the assumptions of linearity, constant error variance, and independence, but does not require normality (see, e.g., Fox, 1984, pp. 42-43). Although the restriction to linear estimators produces simple sampling properties, it is not compelling in light of the vulnerability of least squares to heavy-tailed error distributions.

Second, highly skewed error distributions, aside from their propensity to generate outliers in the direction of the skew, compromise the interpretation of the least-squares fit. This fit is, after all, a conditional mean (of y given the xs), and the mean is not a good measure of the center of a highly skewed distribution. Consequently, we may prefer to transform the data to produce a symmetric error distribution.

Finally, a multimodal error distribution suggests the omission of one or more qualitative variables that divide the data naturally into groups. An examination of the distribution of residuals may therefore motivate respecification of the model.

Although there are tests for non-normal errors, I shall describe here instead graphical methods for examining the distribution of the residuals (but see Chapter 9). These methods are more useful for pinpointing the character of a problem and for suggesting solutions.

Normal Quantile-Comparison Plot of Residuals

One such graphical display is the quantile-comparison plot, which permits us to compare visually the cumulative distribution of an independent random sample—here of studentized residuals—to a cumulative reference distribution—the unit-normal distribution. Note that approximations are implied, because the studentized residuals are t distributed and dependent, but generally the distortion is negligible, at least for moderate-sized to large samples.

To construct the quantile-comparison plot:

1. Arrange the studentized residuals in ascending order: $t_{(1)}, t_{(2)}, \ldots, t_{(n)}$. By convention, the ith largest studentized residual, $t_{(i)}$, has $g_i = (i - 1/2)/n$ proportion of the data below it. This convention avoids cumulative proportions of zero and one by (in effect) counting half of each observation below and half above its recorded value. Cumulative proportions of zero and one would be problematic because the normal distribution, to which we wish to compare the distribution of the residuals, never quite reaches cumulative probabilities of zero or one.
2. Find the quantile of the unit-normal distribution that corresponds to a cumulative probability of g_i—that is, the value z_i from $Z \sim N(0, 1)$ for which $\Pr(Z < z_i) = g_i$.
3. Plot the $t_{(i)}$ against the z_i.

If the t_i were drawn from a unit-normal distribution, then, within the bounds of sampling error, $t_{(i)} = z_i$. Consequently, we expect to find an approximately linear plot with zero intercept and unit slope, a line that can be placed on the plot for comparison. Nonlinearity in the plot, in contrast, is symptomatic of non-normality.

It is sometimes advantageous to adjust the fitted line for the observed center and spread of the residuals. To understand how the adjustment may be accomplished, suppose more generally that a variable X is normally distributed with mean μ and variance ζ^2. Then, for an ordered sample of values, approximately $x_{(i)} = \mu + \zeta z_i$, where z_i is defined as before. In applications, we need to estimate μ and ζ, preferably robustly, because the usual estimators—the sample mean and standard deviation—are markedly affected by extreme values. Generally effective choices are the median of x to estimate μ and $(Q_3 - Q_1)/1.349$ to estimate ζ, where Q_1 and Q_3 are, respectively, the first and third quartiles of x: The median and quartiles are not sensitive to outliers. Note that 1.349 is the number of standard deviations separating the quartiles of a normal distribution. Applied to the studentized residuals, we have the fitted line $\hat{t}_{(i)} = \text{median}(t) + \{[Q_3(t) - Q_1(t)]/1.349\} \times z_i$. The normal quantile-comparison plots in this monograph employ the more general procedure.

Several illustrative normal-probability plots for simulated data are shown in Figure 5.1. In parts a and b of the figure, independent samples of size $n = 25$ and $n = 100$, respectively, were drawn from a unit-normal distribution. In parts c and d, samples of size $n = 100$ were drawn from the highly positively skewed χ^2_4 distribution and the heavy-tailed t_2 distribution, respectively. Note how the skew and heavy tails show up as departures from linearity in the normal quantile-comparison plots. Outliers are discernible as unusually large or small values in comparison with corresponding normal quantiles.

Judging departures from normality can be assisted by plotting information about sampling variation. If the studentized residuals were drawn independently from a unit-normal distribution, then

$$\text{SE}(t_{(i)}) = \frac{1}{\varphi(z_i)} \sqrt{\frac{g_i(1 - g_i)}{n}}$$

where $\varphi(z_i)$ is the probability density (i.e., the "height") of the unit-normal distribution at $Z = z_i$. Thus, $z_i \pm 2 \times \text{SE}(t_{(i)})$ gives a rough 95% confidence interval around the fitted line $\hat{t}_{(i)} = z_i$ in the quantile-

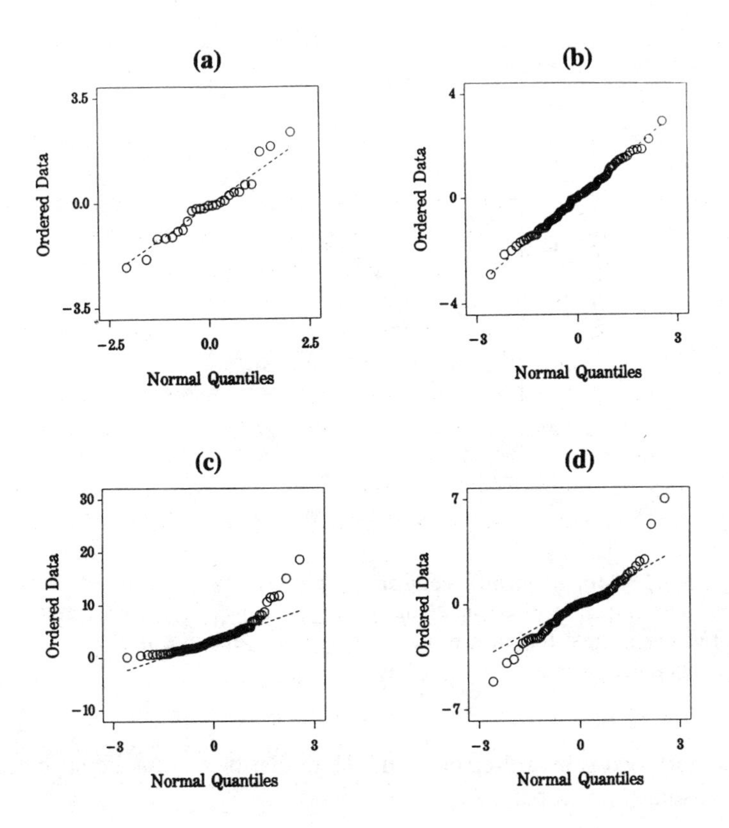

Figure 5.1. Illustrative normal quantile-comparison plots. (a) For a sample of $n = 25$ from $N(0,1)$. (b) For a sample of $n = 100$ from $N(0,1)$. (c) For a sample of $n = 100$ from the positively skewed χ^2_4. (*d*) For a sample of $n = 100$ from the heavy-tailed t_2 .

comparison plot. If the slope of the fitted line is taken as $\hat{\zeta} = (Q_3 - Q_1)/1.349$ rather than 1, then the estimated standard error may be multiplied by $\hat{\zeta}$. As an alternative to computing standard errors, Atkinson (1985) has suggested a computationally intensive simulation procedure that does not treat the studentized residuals as independent and normally distributed.

Figure 5.2 shows a normal quantile-comparison plot for the studentized residuals from Duncan's regression of rated prestige on occupational income and education levels. The plot includes a fitted

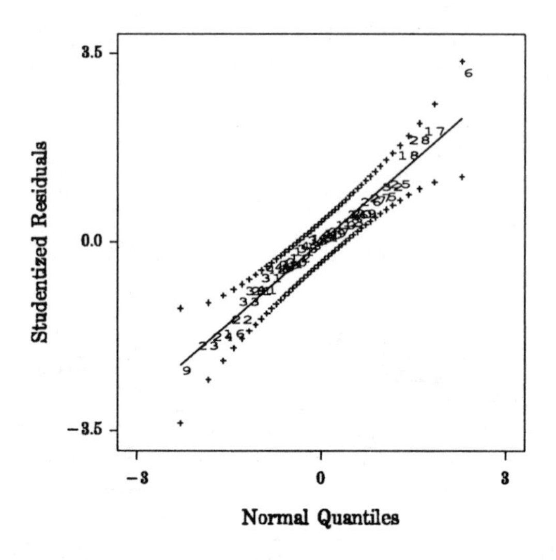

Figure 5.2. Normal quantile-comparison plot for the studentized residuals from the regression of occupational prestige on income and education. The plot shows a fitted line, based on the median and quartiles of the *t*s, and approximate ±2SE limits around the line.

line with two-standard-error limits. Note that the residual distribution is reasonably well behaved.

Histograms of Residuals

A strength of the normal quantile-comparison plot is that it retains high resolution in the tails of the distribution, where problems often manifest themselves. A weakness of the display, however, is that it does not convey a good overall sense of the shape of the distribution of the residuals. For example, multiple modes are difficult to discern in a quantile-comparison plot.

Histograms (frequency bar graphs), in contrast, have poor resolution in the tails or wherever data are sparse, but do a good job of conveying general distributional information. The arbitrary class boundaries, arbitrary intervals, and roughness of histograms sometimes produce misleading impressions of the data, however. These problems can partly be addressed by smoothing the histogram (see Silverman, 1986, or

```
-2*  | 3
-1.  | 977
-1*  | 41
-0.  | 99865
-0*  | 444433110
 0*  | 000011133334
 0.  | 5577788
 1*  | 00
 1.  | 68
 2*  | 0
 2.  |
 3*  | 1
```

Figure 5.3. Stem-and-leaf display of studentized residuals from the regression of occupational prestige on income and education.

Fox, 1990). Generally, I prefer to employ *stem-and-leaf displays*—a type of histogram (Tukey, 1977) that records the numerical data values directly in the bars of the graph—for small samples (say $n < 100$), *smoothed histograms* for moderate-sized samples (say $100 \le n \le 1{,}000$), and histograms with relatively narrow bars for large samples (say $n > 1{,}000$).

A stem-and-leaf display of studentized residuals from the Duncan regression is shown in Figure 5.3. The display reveals nothing of note: There is a single node, the distribution appears reasonably symmetric, and there are no obvious outliers, although the largest value (3.1) is somewhat separated from the next-largest value (2.0).

Each data value in the stem-and-leaf display is broken into two parts: The leading digits comprise the stem; the first trailing digit forms the leaf; and the remaining trailing digits are discarded, thus truncating rather than rounding the data value. (Truncation makes it simpler to locate values in a list or table.) For studentized residuals, it is usually sensible to make this break at the decimal point. For example, for the residuals shown in Figure 5.4: 0.3039 → 0 |3; 3.1345 → 3 |1; and −0.4981 → −0 |4. Note that each stem digit appears twice, implicitly producing bins of width 0.5. Stems marked with asterisks (e.g., 1*) take leaves 0 — 4; stems marked with periods (e.g., 1.) take leaves 5—9. (For more information about stem-and-leaf displays, see, e.g., Velleman and Hoaglin [1981] or Fox [1990].)

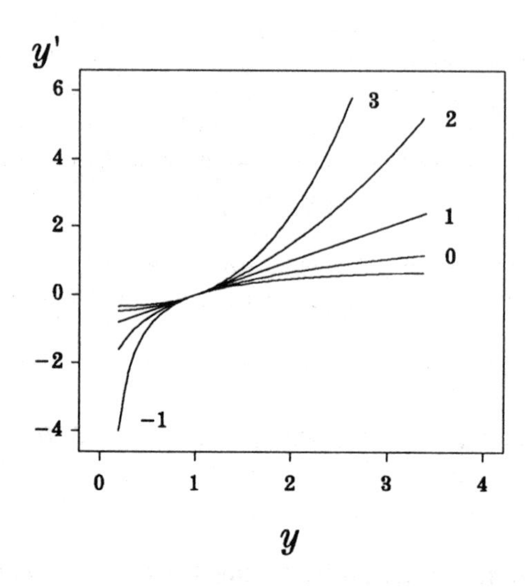

Figure 5.4. The family of powers and roots. The transformation labeled "p" is actually $y' = (y^p - 1)/p$; for $p = 0$, $y' = \log_e y$.

SOURCE: Adapted with permission from Figure 4-1 from Hoaglin, Mosteller, and Tukey (eds.), *Understanding Robust and Exploratory Data Analysis*, © 1983 by John Wiley and Sons, Inc.

Correcting Asymmetry by Transformation

A frequently effective approach to a variety of problems in regression analysis is to transform the data so that they conform more closely to the assumptions of the linear model. In this and later chapters I shall introduce transformations to produce symmetry in the error distribution, to stabilize error variance, and to make the relationship between y and the xs linear.

In each of these cases, we shall employ the family of powers and roots, replacing a variable y (used here generically, because later we shall want to transform xs as well) by $y' = y^p$. Typically, $p = -2, -1, -1/2, 1/2, 2$, or 3, although sometimes other powers and roots are considered. Note that $p = 1$ represents no transformation. In place of the 0th power, which would be useless because $y^0 = 1$ regardless of the value of y, we take $y' = \log y$, usually using base 2 or 10 for the log function. Because logs to different bases differ only by a constant factor, we can select the base for convenience of interpretation. Using the log transformation as a "zeroth power" is reasonable, because the closer p gets to zero, the more y^p looks like the log function (formally, $\lim_{p \to 0}[(y^p - 1)/p] = \log_e y$, where the log to the base $e \approx 2.718$ is the so-called "natural" logarithm). Finally, for negative powers, we take $y' = -y^p$, preserving the order of the y values, which would otherwise be reversed.

As we move away from $p = 1$ in either direction, the transformations get stronger, as illustrated in Figure 5.4. The effect of some of these transformations is shown in Table 5.1a. Transformations "up the ladder" of powers and roots (a term borrowed from Tukey, 1977)—that is, toward y^2—serve differentially to spread out large values of y relative to small ones; transformations "down the ladder"—toward log y—have the opposite effect. To correct a positive skew (as in Table 5.1b), it is therefore necessary to move down the ladder; to correct a negative skew (Table 5.1c), which is less common in applications, move up the ladder.

I have implicitly assumed that all data values are positive, a condition that must hold for power transformations to maintain order. In practice, negative values can be eliminated prior to transformation by adding a small constant, sometimes called a "start," to the data. Likewise, for power transformations to be effective, the ratio of the largest to the smallest data value must be sufficiently large; otherwise the transformation will be too nearly linear. A small ratio can be dealt with by using a negative start.

In the specific context of regression analysis, a skewed error distribution, revealed by examining the distribution of the residuals, can often be corrected by transforming the dependent variable. Although more sophisticated approaches are available (see, e.g., Chapter 9), a good transformation can be located by trial and error.

Dependent variables that are bounded below, and hence that tend to be positively skewed, often respond well to transformations down the ladder of powers. Power transformations usually do not work well, however, when many values stack up against the boundary, a situation termed *truncation* or *censoring* (see, e.g., Tobin [1958] for a treatment of "limited" dependent variables in regression). As well, data that are bounded both

TABLE 5.1

Correcting Skews by Power Transformations

(a) Effect of power transformations on the spacing of scores.

$-1/y$	$\log_{10} y$	←	y	→	y^2	y^3
−1	0		1		1	1
}1/2[a]	}0.30		}1		}3	}7
−1/2	0.30		2		4	8
}1/6	}0.18		}1		}5	}19
−1/3	0.48		3		9	27
}1/12	}0.12		}1		}7	}37
−1/4	0.60		4		16	64
}1/20	}0.10		}1		}9	}61
−1/5	0.70		5		25	125

(b) Descending the ladder of powers to correct a positive skew, pulling in the right tail.

y	→	$\log_{10} y$
1		0
}9		}1
10		1
}90		}1
100		2
}900		}1
1000		3

(c) Ascending the ladder of power to correct a negative skew, pulling in the left tail.

y	→	y^2
1.000		1
}0.414		}1
1.414		2
}0.318		}1
1.732		3
}0.268		}1
2.000		4

a. The interlinear numbers give the differences between adjacent scores.

above and below—such as proportions and percentages—generally require another approach. For example the logit or "log odds" transformation given by $y' = \log[y/(1 - y)]$, often works well for proportions.

Transforming variables in a regression analysis raises issues of interpretation. I address these issues briefly at the end of Chapter 7.

6. NONCONSTANT ERROR VARIANCE

Detecting Nonconstant Error Variance

One of the assumptions of the regression model is that the variation of the dependent variable around the regression surface—the error variance—is everywhere the same: $V(\varepsilon) = V(y \mid x_1, \ldots, x_k) = \sigma^2$. Nonconstant error variance is often termed "heteroscedasticity." Although the least-squares estimator is unbiased and consistent even when the error variance is not constant, its efficiency is impaired and the usual formulas for coefficient standard errors are inaccurate, the degree of the problem depending on the degree to which error variances differ. I describe graphical methods for detecting nonconstant error variance in this chapter. Tests for heteroscedasticity are discussed in Chapter 8 on discrete data and in Chapter 9 on maximum-likelihood methods.

Because the regression surface is k dimensional, and imbedded in a space of $k + 1$ dimensions, it is generally impractical to assess the assumption of constant error variance by direct graphical examination of the data for k larger than 1 or 2. Nevertheless, it is common for error variance to increase as the expectation of y grows larger, or there may be a systematic relationship between error variance and a particular x. The former situation can be detected by plotting residuals against fitted values, and the latter by plotting residuals against each x. It is worth noting that plotting residuals against y (as opposed to $\hat{y}$) is generally unsatisfactory. The plot will be tilted: There is a built-in linear correlation between e and y, because $y = \hat{y} + e$; in fact, the correlation between y and e is $r(y, e) = \sqrt{1 - R^2}$. In contrast, the least-squares fit insures that $r(\hat{y}, e) = 0$, producing a plot that is much easier to examine for evidence of nonconstant spread.

Because the least-squares residuals have unequal variances even when the errors have constant variance, I prefer plotting the studentized residuals against fitted values. Finally, a pattern of changing spread is often more easily discerned in a plot of $|t_i|$ or t_i^2 versus $\hat{y}$, perhaps augmented by a lowess scatterplot smooth (see Appendix A6.1); smoothing this plot is particularly useful when the sample size is very large or the distribution of $\hat{y}$ is very uneven. An example appears in Figure 6.2.

An illustrative plot of studentized residuals against fitted values is shown in Figure 6.1a. In Figure 6.1b, studentized residuals are plotted

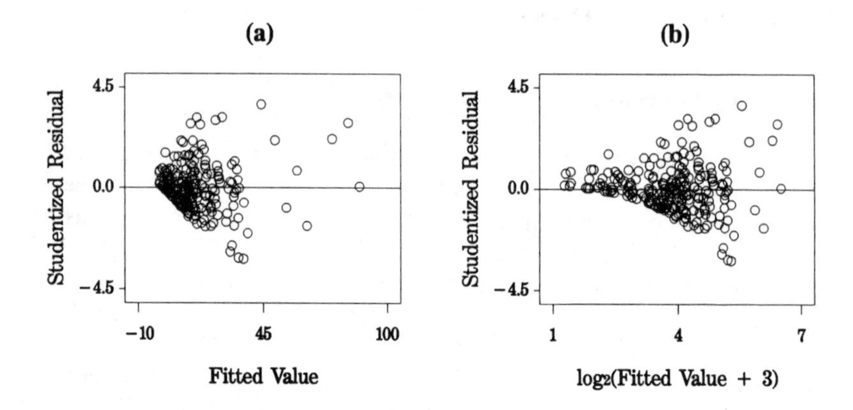

Figure 6.1. Plots of studentized residuals versus fitted values for Ornstein's interlocking-directorate regression. (a) t versus $\hat{y}$. (b) t versus $\log_2(3 + \hat{y})$. The log transformation serves to reduce the skew of the fitted values, making the increasing residual spread easier to discern.

against $\log_2(3 + \hat{y})$; by correcting the positive skew in these $\hat{y}$ values, the second plot makes it easier to discern the tendency of the residual spread to increase with $\hat{y}$. The data for this example are drawn from work by Ornstein (1976) on interlocking directorates among the 248 largest Canadian firms. The number of interlocking directorate and executive positions maintained by the corporations is regressed on corporate assets (square-root transformed to make the relationship linear; see Chapter 7); 9 dummy variables representing 10 industrial classes, with heavy manufacturing serving as the baseline category; and 3 dummy variables representing 4 nations of control, with Canada as the baseline category. The results of the regression are given in the left-hand columns of Table 6.1; the results shown on the right of the table are discussed below. Note that part of the tendency for the residual scatter to increase with $\hat{y}$ is due to the lower bound of 0 for y: Because $e = y - \hat{y}$, the smallest possible residual corresponding to a particular $\hat{y}$ value is $e = 0 - \hat{y} = -\hat{y}$.

Correcting Nonconstant Error Variance

Transformations frequently serve to correct a tendency of the error variance to increase or, less commonly, decrease with the magnitude

TABLE 6.1

Regression of Number of Interlocking Directorate and Executive Positions Maintained by 248 Major Canadian Corporations on Corporate Assets, Sector, and Nation of Control

	Interlocks		$\sqrt{Interlocks + 1}$	
Regressor	*Coefficient*	*SE*	*Coefficient*	*SE*
Constant	4.19	1.85	2.33	0.231
$\sqrt{\text{Assets}}$	0.252	0.019	0.0260	0.00232
Sector[a]				
Agriculture, food, light industry	−1.20	2.04	−0.0567	0.255
Mining, metals	0.342	2.01	0.356	0.252
Wood, paper	5.15	2.68	0.786	0.335
Construction	−5.13	4.70	−0.740	0.588
Transport	−0.381	2.82	0.354	0.353
Merchandizing	−0.867	2.63	0.148	0.329
Banking	−14.4	5.58	−2.25	0.697
Other financials	−5.70	2.93	−0.0880	0.366
Holding companies	−2.43	4.01	−0.245	0.502
Nation of Control[b]				
United States	−8.09	1.48	−1.11	0.185
Britain	−4.44	2.65	−0.527	0.331
Other	−1.16	2.66	−0.114	0.333
R^2	0.655		0.580	

SOURCE: Data taken from M. Ornstein, personal communication; the data appear in Fox (1984).
a. Baseline category for zero/one dummy variables: Heavy manufacturing.
b. Baseline category: Canada.

of the dependent variable: Move y down the ladder of powers and roots if the residual scatter broadens with the fitted values; move y up the ladder if the residual scatter narrows. An effective transformation may be selected by trial and error (but see Chapter 9 for an analytic method of selecting a variance-stabilizing transformation).

If the error variance is proportional to a particular x, or if the pattern of $V(\varepsilon_i)$ is otherwise known up to a constant of proportionality, then an alternative to transformation of y is weighted-least-squares (WLS) estimation (see Appendix A6.2). It also is possible to correct the estimated standard errors of the least-squares coefficients for heteroscedasticity: A method proposed by White (1980) is described

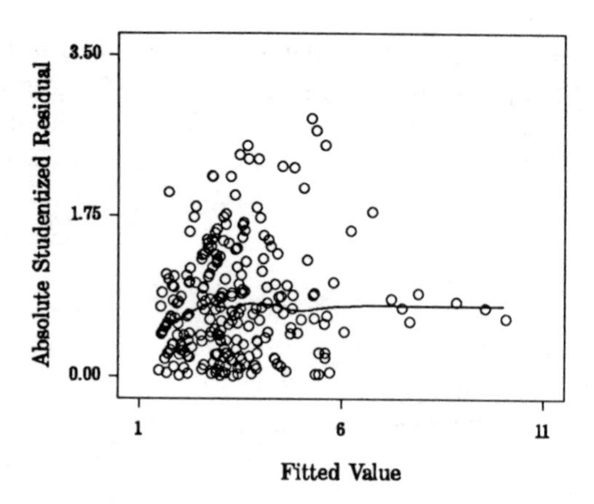

Figure 6.2. Plot of absolute studentized residuals versus fitted values for the square-root transformed interlocking-directorate data. The line on the graph is a lowess smooth using $f = 0.5$ and 2 robustness iterations.

in Appendix A6.3. An advantage of this approach is that knowledge of the *pattern* of nonconstant error variance (e.g., increased variance with the level of y or with an x) is not required. If, however, the heteroscedasticity problem is severe, and the corrected standard errors therefore are substantially larger than those produced by the usual formula, then discovering the pattern of nonconstant variance and correcting for it—by a transformation or WLS estimation—offers the possibility of more efficient estimation. In any event, unequal error variances are worth correcting only when the problem is extreme—where, for example, the spread of the errors varies by a factor of about three or more (i.e., the error variance varies by a factor of about 10 or more; see Appendix A6.4).

For Ornstein's interlocking-directorate regression, for example, a square-root transformation appears to correct the dependence of the residual spread on the level of the dependent variable. A plot of $|t_i|$ versus $\hat{y}_i$ for the transformed data is given in Figure 6.2, and the regression results appear in the right-hand columns of Table 6.1. The lowess smooth in Figure 6.2 (see Appendix A6.1) shows little change in the average absolute studentized residuals as the fitted values increase.

The coefficients for the original and transformed regressions in Table 6.1 cannot be compared directly, because the scale of the dependent variable has been altered. It is clear, however, that assets retain their positive effect and that the nations of control maintain their ranks. The sectoral ranks are also similar across the two analyses, although not identical. In comparing the two sets of results, recall that the baseline categories for the sets of dummy regressors—Canada and heavy manufacturing—implicitly have coefficients of zero.

Transforming y also changes the shape of the error distribution and alters the shape of the regression of y on the xs. It is frequently the case that producing constant residual variation through a transformation also makes the distribution of the residuals more symmetric. At times, eliminating nonconstant spread also makes the relationship of y to the xs more nearly linear (see the next chapter). These by-products are not necessary consequences of correcting nonconstant error variance, however, and it is particularly important to check data for nonlinearity following a transformation of y. Of course, because there generally is no reason to suppose that the regression is linear prior to transforming y, we should check for nonlinearity even when y is *not* transformed.

Finally, nonconstant residual spread sometimes is evidence for the omission of important effects from the model. Suppose, for example, that there is an omitted categorical independent variable, such as regional location, that interacts with assets in affecting interlocks; in particular, the assets slope, although positive in every region, is steeper in some regions than in others. Then the omission of region and its interactions with assets could produce a fan-shaped residual plot even if the errors from the correct model have constant spread. The detection of this type of specification error therefore requires substantive insight into the process generating the data and cannot rely on diagnostics alone.

7. NONLINEARITY

The assumption that $E(\varepsilon)$ is everywhere zero implies that the specified regression surface captures the dependency of y on the xs. Violating the assumption of linearity therefore implies that the model fails to capture the systematic pattern of relationship between the dependent and independent variables. For example, a partial relationship specified to be linear may be nonlinear, or two independent variables specified to have additive partial effects may interact in determining y. Nevertheless, the fitted model is frequently a useful approximation even if the regression surface $E(y)$ is not precisely captured. In other instances, however, the model can be extremely misleading.

The regression surface is generally high dimensional, even after accounting for regressors (such as polynomial terms, dummy variables, and interactions) that are functions of a smaller number of fundamental independent variables. As in the case of nonconstant error variance, therefore, it is necessary to focus on particular patterns of departure from linearity. The graphical diagnostics discussed in this chapter represent two-dimensional views of the higher-dimensional point-cloud of observations $\{y_i, x_{1i}, \ldots, x_{ki}\}$. With modern computer graphics, the ideas here can usefully be extended to three dimensions, permitting, for example, the detection of two-way interactions between independent variables (Monette, 1990).

Residual and Partial-Residual Plots

Although it is useful in multiple regression to plot y against each x, these plots do not tell the whole story—and can be misleading—because our interest centers on the *partial* relationship between y and each x, controlling for the other xs, not on the *marginal* relationship between y and a single x. Residual-based plots are consequently more relevant in this context.

Plotting residuals or studentized residuals against each x, perhaps augmented by a lowess smooth (see Appendix A6.1), is helpful for detecting departures from linearity. As Figure 7.1 illustrates, however, residual plots cannot distinguish between monotone (i.e., strictly increasing or decreasing) and nonmonotone (e.g., falling and then rising) nonlinearity. The distinction between monotone and nonmonotone nonlinearity is lost in the residual plots because the least-squares fit ensures that the residuals are linearly uncorrelated with each x. The distinction is important, because, as I shall explain below, monotone nonlinearity frequently can be corrected by simple transformations. In Figure 7.1, for example, case a might be modeled by $y = \beta_0 + \beta_1 x^2 + \varepsilon$, whereas case b cannot be linearized by a power transformation of x and might instead be dealt with by a quadratic specifi-

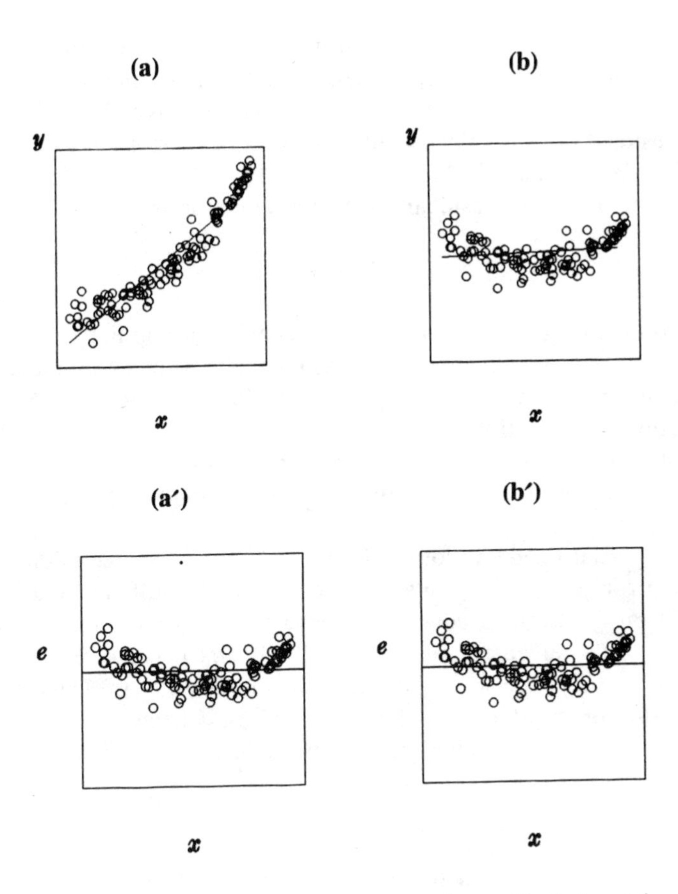

Figure 7.1. Scatterplots (a and b) and corresponding residual plots (a′ and b′) in simple regression. The residual plots do not distinguish between (a) a nonlinear but monotone relationship, and (b) a nonlinear, nonmonotone relationship.

cation, $y = \beta_0 + \beta_1 x + \beta_2 x^2 + \varepsilon$. (Case b could, however, be accommodated by a more complex transformation of x: $y = \beta_0 + \beta_1(x - \alpha)^2 + \varepsilon$; I shall not pursue this approach here.)

In contrast to simple residual plots, partial-regression plots, introduced in Chapter 4 for detecting influential data, can reveal nonlinearity and suggest whether a relationship is monotone. These plots are not always useful for locating a transformation, however: The partial-regression plot adjusts x_j for the other xs, but it is the unadjusted x_j that is transformed in respecifying the model. Partial-residual plots, also called *component-plus-residual plots,* are often an effective alternative. Partial-residual plots are not as suitable as partial-regression plots for revealing leverage and influence.

Define the partial residual for the jth regressor as

$$e_i^{(j)} = e_i + b_j x_{ji}$$

In words, add back the linear component of the partial relationship between y and x_j to the least-squares residuals, which may include an unmodeled nonlinear component. Then plot $e^{(j)}$ versus x_j. By construction, the multiple-regression coefficient b_j is the slope of the simple linear regression of $e^{(j)}$ on x_j, but nonlinearity should be apparent in the plot as well. Again, a lowess smooth may help in interpreting the plot.

The partial-residual plots in Figure 7.2 are for a regression of the rated prestige P of 102 Canadian occupations (from Pineo and Porter, 1967) on the average education (E) in years, average income (I) in dollars, and percentage of women (W) in the occupations in 1971. (Related results appear in Fox and Suschnigg [1989]; cf. Duncan's regression for similar U.S. data reported in Chapter 4.) A lowess smooth is shown in each plot. The results of the regression are as follows:

$$\underset{}{\hat{P}} = \underset{(3.24)}{-6.79} + \underset{(0.39)}{4.19E} + \underset{(0.00028)}{0.00131I} - \underset{(0.0304)}{0.00891W}$$

$$R^2 = 0.80 \qquad s = 7.85$$

Note that the magnitudes of the regression coefficients should not be compared, because the independent variables are measured in different units: In particular, the unit for income is small—the dollar—and that for education is comparatively large—the year. Interpreting the regression coefficients in light of the units of the corresponding independent variables, the education and income coefficients are both substantial, whereas the coefficient for percentage of women is very small.

There is apparent monotone nonlinearity in the partial-residual plots for education and, much more strongly, income (Figure 7.2, parts a and b); there also is a small apparent tendency for occupations

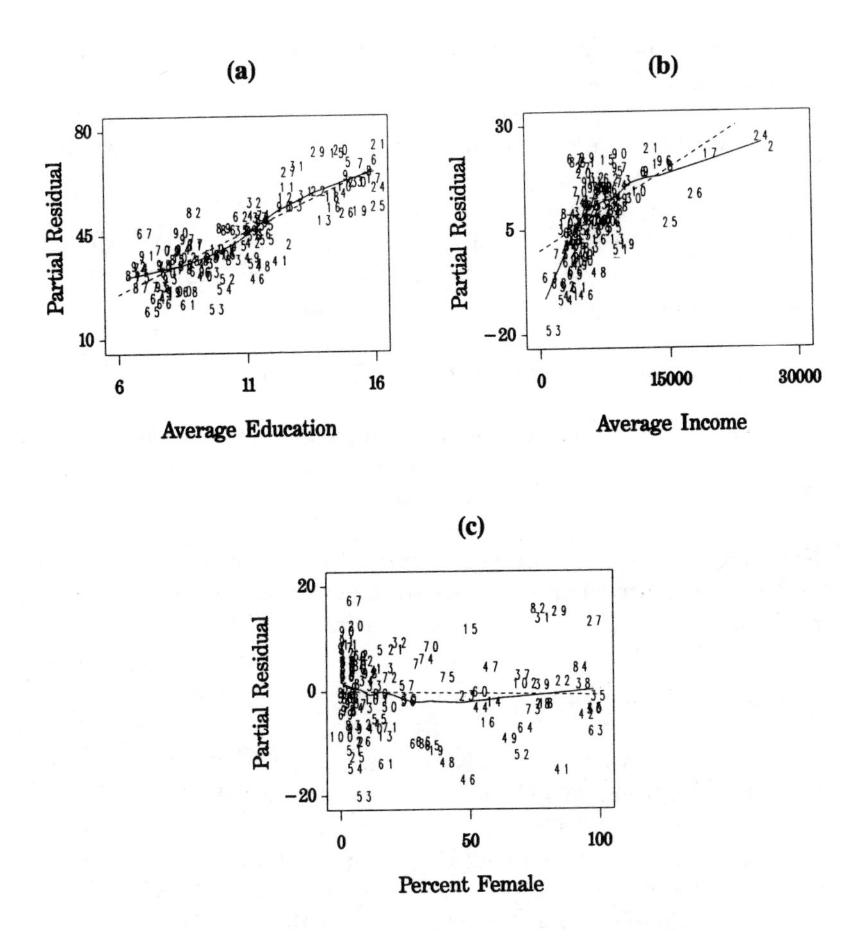

Figure 7.2. Partial-residual plots for the regression of the rated prestige of 102 Canadian occupations on 1971 occupational characteristics: (a) education, (b) income, and (c) percentage of women. The observation index is plotted for each point. In each graph, the linear least-squares fit (broken line) and the lowess smooth (solid line for $f = 0.5$ with 2 robustness iterations) are shown.

SOURCE: Data taken from B. Blishen, W. Carroll, and C. Moore, personal communication; *Census of Canada* (Statistics Canada, 1971, Part 6, pp. 19.1-19.21); Pineo and Porter (1967).

with intermediate percentages of women to have lower prestige, controlling for income and educational levels (Figure 7.2c). To my eye, the patterns in the partial-residual plots for education and percentage of women are not easily discernible without the lowess smooth: The departure from linearity is not great. The nonlinear patterns for income and percentage of women are simple: In the first case, the lowess curve opens downwards; in the second case, it opens upwards. For education, however, the direction of curvature changes, producing a more complex nonlinear pattern.

Mallows (1986) has suggested a variation on the partial-residual plot that sometimes reveals nonlinearity more clearly: First, add a quadratic term in x_j to the model, which becomes

$$y_i = \beta_0 + \beta_1 x_{1i} + \ldots + \beta_j x_{ji} + \gamma x_{ji}^2 + \ldots + \beta_k x_{ki} + \varepsilon_i$$

Then, after fitting the model, form the "augmented" partial residual

$$e_i'^{(j)} = e_i + b_j x_{ji} + c x_{ji}^2$$

Note that in general b_j differs from the regression coefficient for x_j in the original model, which does not include the squared term. Finally, plot $e'^{(j)}$ versus x_j.

Transformations for Linearity

To consider how power transformations can serve to linearize a monotone nonlinear relationship, examine Figure 7.3. Here, I have plotted $y = (1/5)x^2$ for $x = 1, 2, 3, 4, 5$. By construction, the relationship can be linearized by taking $x' = x^2$, in which case $y = (1/5)x'$; or by taking $y' = \sqrt{y}$, in which case $y' = \sqrt{1/5}\ x$. Figure 7.3 reveals how each transformation serves to stretch one of the axes differentially, pulling the curve into a straight line.

As illustrated in Figure 7.4, there are four simple patterns of monotone nonlinear relationships. Each can be straightened by moving y, x, or both up or down the ladder of powers and roots: The direction of curvature determines the direction of movement on the ladder; Tukey (1977) calls this the "bulging rule." Specific transformations to linearity can be located by trial and error (but see Chapter 9 for an analytic approach).

In multiple regression, the bulging rule may be applied to the partial-residual plots. Generally, we transform x_j in preference to y, because changing the scale of y disturbs its relationship to other

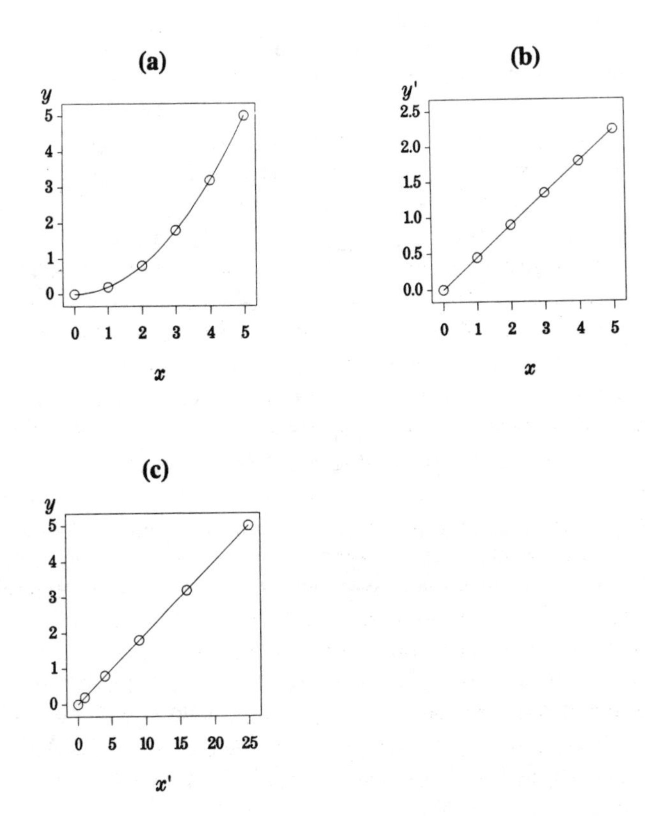

Figure 7.3. How a transformation of y (a to b) or x (a to c) can make a simple monotone nonlinear relationship linear.

regressors and because transforming y changes the error distribution. An exception occurs when similar nonlinear patterns are observed in all of the partial-residual plots. Furthermore, the logit transformation often helps for dependent variables that are proportions.

As suggested in connection with Figure 7.1b, nonmonotone nonlinearity (and some complex monotone patterns) frequently can be accommodated by fitting polynomial functions in an x; quadratic specifications are often useful in applications. As long as the model remains linear in its parameters, it may be fit by linear least-squares regression.

Trial-and-error experimentation with the Canadian occupational prestige data leads to the log transformation of income. The possibly

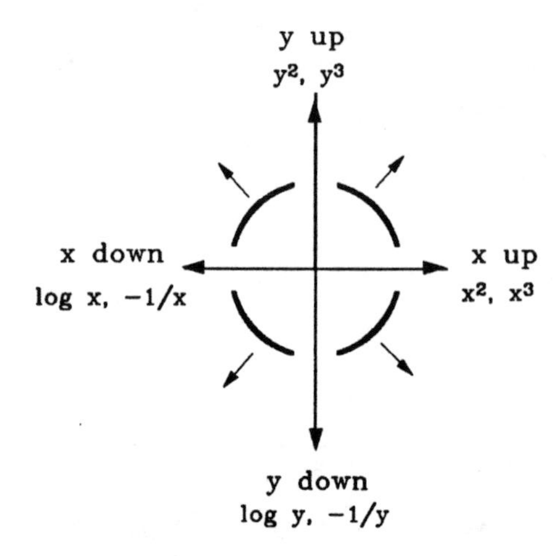

Figure 7.4. Determining a transformation to linearity by the "bulging rule."
SOURCE: Adapted from Tukey, *Exploratory Data Analysis*, © 1977 by Addison-Wesley Publishing Co. Adapted and reprinted by permission of Addison-Wesley Publishing Co., Inc., Reading, MA.

curvilinear partial relationship of prestige to the percentage of women in the occupations suggests the inclusion of linear and quadratic terms for this independent variable. These changes produce a modest, though discernible, improvement in the fit of the model:

$$\hat{P} = -111 + 3.77E + 9.36\log_2 I - 0.139W + 0.00215W^2$$
$$(15) \quad (0.35) \quad (1.30) \qquad (0.087) \quad (0.00094)$$

$$R^2 = 0.84 \qquad s = 6.95$$

Note the statistically significant quadratic term for percentage of women. The partial effect of this variable is relatively small, however, ranging from a minimum of −2.2 prestige points for an occupation with 32% women to 7.6 points for a hypothetical occupation consisting entirely of women. Because the nonlinear pattern in the partial-residual plot for education is complex, a power transformation of this independent variable is not promising: Trial and error suggests that the best that we can do is to increase R^2 to 0.85 by squaring education.

In transforming data or respecifying the functional form of the model, there should be an interplay between substantive and modeling considerations. We must recognize, however, that social theories are almost never mathematically concrete: Theory may tell us that prestige should increase with income, but it does not specify the functional form of the relationship.

Still, in certain contexts, specific transformations may have advantages of interpretability. For example, log transformations often can be given meaningful substantive interpretation: To increase $\log_2 x$ by 1, for instance, represents a doubling of x. In the respecified Canadian occupational prestige regression, therefore, doubling income is associated on average with a 9-point increment in prestige, holding education and gender composition constant.

Likewise, the square root of an area or cube root of a volume can be interpreted as a linear measure of distance or length, the inverse of the amount of time required to traverse a particular distance is speed, and so on. If *both* y and x_j are log-transformed, then the regression coefficient for x_j' is interpretable as the "elasticity" of y with respect to x_j—that is, the approximate percentage of change in y corresponding to a 1% change in x_j. In many contexts, a quadratic relationship will have a clear substantive interpretation (in the example, occupations with a gender mix appear to pay a small penalty in prestige), but a fourth-degree polynomial may not.

Finally, although it is desirable to maintain simplicity and interpretability, it is not reasonable to distort the data by insisting on a functional form that is clearly inadequate. It is possible, in any event, to display the fitted relationship between y and an x graphically or in a table, using the original scales of the variables if they have been transformed, or to describe the effect at a few strategic x values (see, e.g., the brief description above of the partial effect of percentage of women on occupational prestige).

8. DISCRETE DATA

Discrete independent and dependent variables often lead to plots that are difficult to interpret. A simple example of this phenomenon appears in Figure 8.1, the data for which are drawn from the 1989 *General Social Survey* conducted by the National Opinion Research Center. The independent variable, years of education completed, is coded from 0 to 20. The dependent variable is the number of correct

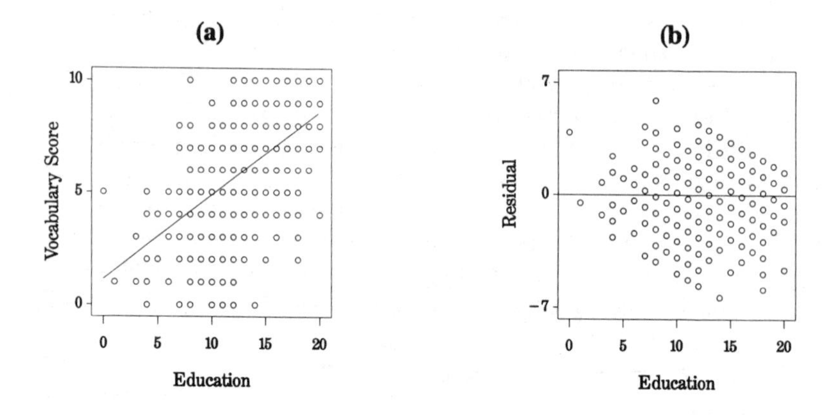

Figure 8.1. Scatterplot (a) and residual plot (b) for vocabulary score by year of education. The least-squares regression line is shown on the scatterplot.

answers to a 10-item vocabulary test; note that this variable is a disguised proportion—literally, the proportion correct × 10.

The scatterplot in Figure 8.1a conveys the general impression that vocabulary increases with education. The plot is difficult to read, however, because most of the 968 data points fall on top of one another. The least-squares regression line, also shown on the plot, has the equation

$$\hat{V} = \underset{(0.28)}{1.13} + \underset{(0.021)}{0.374}E$$

$$R^2 = 0.248 \qquad s = 1.92$$

where V and E are, respectively, the vocabulary score and education.

Figure 8.1b plots residuals from the fitted regression against education. The diagonal lines running from upper left to lower right in this plot are typical of residuals for a discrete dependent variable: For any one of the 11 distinct y values, e.g., $y = 5$, the residual is $e = 5 - b_0 - b_1x = 3.87 - 0.374x$, which is a linear function of x. I noted a similar phenomenon in Chapter 6 for the plot of residuals against fitted values when y has a fixed minimum score. The diagonals from lower left to upper right are due to the discreteness of x.

It also appears that the variation of the residuals in Figure 8.1b is lower for the largest and smallest values of education than for intermediate values. This pattern is consistent with the observation that the dependent variable is a disguised proportion: As the average number of correct answers approaches 0 or 10, the potential variation in vocabulary scores decreases. It is possible, however, that at least part of the apparent decrease in residual variation is due to the relative sparseness of data at the extremes of the education scale. Our eye is drawn to the range of residual values, especially because we cannot see most of the data points, and even when variance is constant, the range tends to increase with the amount of data.

These issues are addressed in Figure 8.2, where each data point has been randomly "jittered" both vertically and horizontally: Specifically, a uniform random variable on the interval [−1/2, 1/2] was added to each education and vocabulary score. This approach to plotting discrete data was suggested by Chambers, Cleveland, Kleiner, and Tukey (1983). The plot also shows the fitted regression line for the original data, along with lines tracing the median and first and third quartiles of the distribution of jittered vocabulary scores for each value of education; I excluded education values below six from the median and quartile traces because of the sparseness of data in this region.

Several features of Figure 8.2 are worth highlighting: (a) It is clear from the jittered data that the observations are particularly dense at 12 years of education, corresponding to high-school graduation; (b) the median trace is quite close to the linear least-squares regression line; and (c) the quartile traces indicate that the spread of y does not decrease appreciably at high values of education.

A discrete *dependent* variable violates the assumption that the error in the regression model is normally distributed with constant variance. This problem, like that of a limited dependent variable, is only serious in extreme cases—for example, when there are very few response categories, or where a large proportion of observations is in a small number of categories, conditional on the values of the independent variables.

In contrast, discrete *independent* variables are perfectly consistent with the regression model, which makes no distributional assumptions about the xs other than uncorrelation with the error. Indeed a discrete x makes possible a straightforward hypothesis test of nonlinearity, sometimes called a test for "lack of fit." Likewise, it is relatively

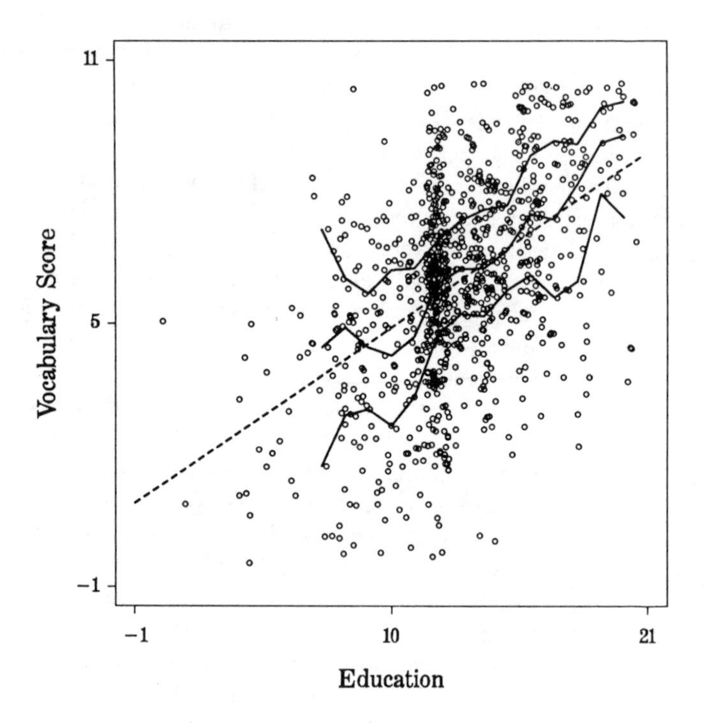

Figure 8.2. "Jittered" scatterplot for vocabulary score by education. A small random quantity is added to each horizontal and vertical coordinate. The dashed line is the least-squares regression line for the unjittered data. The solid lines are median and quartile traces for the jittered vocabulary scores.

simple to test for nonconstant error variance across categories of a discrete independent variable (see below).

Testing for Nonlinearity

Suppose, for example, that we model education with a set of dummy regressors rather than specify a linear relationship between vocabulary score and education. Although there are 21 conceivable education scores, ranging from 0 through 20, none of the individuals in the sample has 2 years of education, yielding 20 categories and 19 dummy regressors. The model becomes

TABLE 8.1
Analysis of Variance for Vocabulary-Test Score, Showing the Incremental F Test for Nonlinearity of the Relationship Between Vocabulary and Education

Source	*df*	*Sum of Squares*	*Mean Square*	*F*	*p*
Education (Model 8.1)	19	1262.0	66.40	18.1	<<0.0001
Linear (Model 8.2)	1	1175.0	1175.0	320.0	<<0.0001
Nonlinear ("lack of fit")	18	86.58	4.810	1.31	0.17
Error ("pure error")	948	3473.0	3.663		
Total	967	4735.0			

SOURCE: Data taken from 1989 General Social Survey, National Opinion Research Center.

$$y_i = \gamma_0 + \gamma_1 d_{1i} + \ldots + \gamma_{19} d_{19,i} + \varepsilon_i \qquad [8.1]$$

Contrasting this model with

$$y_i = \beta_0 + \beta_1 x_i + \varepsilon_i \qquad [8.2]$$

produces a test for nonlinearity, because Equation 8.2, specifying a linear relationship, is a special case of Equation 8.1, which captures any pattern of relationship between $E(y)$ and x. The resulting incremental F test for nonlinearity appears in the analysis-of-variance of Table 8.1. There is, therefore, very strong evidence of a linear relationship between vocabulary and education, but little evidence of nonlinearity.

The F test for nonlinearity easily can be extended to a discrete independent variable—say, x_1—in a multiple-regression model. Here, we contrast the more general model

$$y = \gamma_0 + \gamma_1 d_1 + \ldots + \gamma_{q-1} d_{q-1} + \beta_2 x_2 + \ldots + \beta_k x_k + \varepsilon$$

with a model specifying a linear effect of x_1,

$$y = \beta_0 + \beta_1 x_1 + \beta_2 x_2 + \ldots + \beta_k x_k + \varepsilon$$

where $d_1, \ldots, d_{q-1}$ are dummy regressors constructed to represent the q categories of x_1.

Testing for Nonconstant Error Variance

A discrete x (or combination of xs) partitions the data into q groups. Let y_{ij} denote the jth of n_i dependent-variable scores in the ith group. If the error variance is constant, then the within-group variance estimates

$$s_i^2 = \frac{\sum_{j=1}^{n_i} (y_{ij} - \bar{y}_i)^2}{n_i - 1}$$

should be similar. Here, $\bar{y}_i$ is the mean in the ith group. Tests that examine the s_i^2 directly, such as Bartlett's (1937) commonly employed test, do not maintain their validity well when the errors are non-normal.

Many alternative tests have been proposed. In a large-scale simulation study, Conover, Johnson, and Johnson (1981) demonstrate that the following simple F test is both robust and powerful: Calculate the values $z_{ij} = |y_{ij} - y_i^*|$ where y_i^* is the median y within the ith group. Then perform a one-way analysis-of-variance of the variable z over the q groups. If the error variance is not constant across the groups, then the group means $\bar{z}_i$ will tend to differ, producing a large value of the F test statistic. For the vocabulary data, for example, where education partitions the 968 observations into $q = 20$ groups, this test gives $F_{19,948} = 1.48$, $p = .08$, providing weak evidence of nonconstant spread.

9. MAXIMUM-LIKELIHOOD METHODS, SCORE TESTS, AND CONSTRUCTED VARIABLES

The methods developed in this chapter rely on maximum-likelihood estimation (see, e.g., Fox, 1984, Appendix C, for a concise summary; Wonnacott and Wonnacott, 1990, Ch. 18, for an elementary introduction). The rationale for these methods is more complex than for parallel ad hoc procedures outlined in earlier sections, but their implementation is nevertheless straightforward. The material in this chapter should therefore prove useful even to data analysts relatively unschooled in the subtleties of the underlying statistical theory.

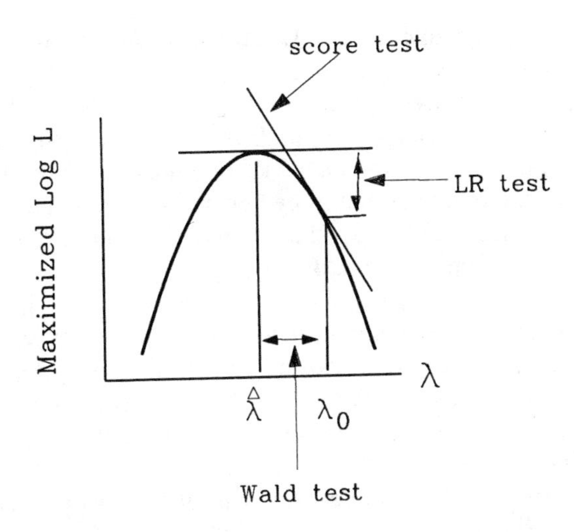

Figure 9.1. The likelihood-ratio, Wald, and score tests for the hypothesis H_0: $\lambda = \lambda_0$.

A statistically sophisticated approach to selecting a transformation of y or an x is to embed the usual multiple-regression model in a more general model that contains a parameter for the transformation. If several variables are to be transformed, or if the transformation is complex, then there may be several such parameters. Models of this type are fundamentally nonlinear.

Suppose that the transformation is indexed by a single parameter λ, and that we can write down the likelihood for the model as a function of the transformation parameter and the usual regression parameters: $L(\lambda, \beta_0, \beta_1, \ldots, \beta_k, \sigma^2)$. Maximizing the likelihood yields the maximum-likelihood estimate (MLE) of λ along with the MLEs of the other parameters. Now suppose that $\lambda = \lambda_0$ represents *no* transformation (e.g., $\lambda_0 = 1$ for the power transformation y^{λ}). The likelihood-ratio test of H_0: $\lambda = \lambda_0$ assesses the evidence that a transformation is required.

As illustrated in Figure 9.1, the likelihood-ratio test compares the log-likelihood at the MLE $\hat{\lambda}$ with the log-likelihood at the null value λ_0: If $\log_e L(\hat{\lambda})$ is sufficiently larger than $\log_e L(\lambda_0)$, then H_0 is rejected, and we conclude that a transformation is required. Alternative tests, also illustrated in Figure 9.1, are the Wald test, based on the distance between $\hat{\lambda}$ and λ_0; and the score test (also called the "Lagrange multiplier" test), based on the slope of the log-likelihood at λ_0—a steep slope casts doubt on H_0, because the log-likelihood is flat at the maximum (i.e., when $\lambda = \hat{\lambda}$). For a quadratic log-likelihood, the three tests are identical, but more generally they are not, although they usually provide similar p values in practice and have equivalent asymptotic (large-sample) properties.

A disadvantage of the likelihood-ratio and Wald tests is that they require finding the MLE, which usually necessitates iteration (a repetitive process of successively closer approximations). The slope of $\log_e L$ at λ_0, in contrast, generally can be assessed or approximated without iteration. Often the score test can be formulated as the t statistic for a new regressor, called a *constructed variable,* to be added to the regression model. Moreover, a partial-regression plot for the constructed variable then can reveal whether one or a small number of observations is unduly influential in determining the transformation or whether evidence for the transformation is spread throughout the data.

Box-Cox Transformation of *y*

Box and Cox (1964) have suggested power transformation of y with the object (as much as possible) of normalizing the error distribution, stabilizing the error variance, and straightening the relationship of y to the xs. The general model is

$$y_i^{(\lambda)} = \beta_0 + \beta_1 x_{1i} + \ldots + \beta_k x_{ki} + \varepsilon_i$$

$$\varepsilon_i \sim \text{NID}(0, \sigma^2)$$

where

$$y_i^{(\lambda)} = \begin{cases} \dfrac{y_i^{\lambda} - 1}{\lambda} & \text{for } \lambda \neq 0 \\ \log_e y_i & \text{for } \lambda = 0 \end{cases}$$

and where all y_i are positive. For a particular choice of λ, Box and Cox show that the conditional maximized log-likelihood is

$$\log_e L(\beta_0, \beta_1, \ldots, \beta_k, \sigma^2 \mid \lambda) = -\frac{n}{2}(1 + \log_e 2\pi)$$

$$-\frac{n}{2}\log_e s^2(\lambda) + (\lambda - 1)\sum_{i=1}^{n} \log_e y_i$$

where $s^2(\lambda) = \Sigma\, e^2_{(\lambda)i}/n$, and where the $e_{(\lambda)i}$ are the residuals from the least-squares linear regression of $y^{(\lambda)}$ on the xs.

A simple procedure for finding the MLE $\hat{\lambda}$, then, is to evaluate the maximized $\log_e L$ for a range of values of λ, say between −2 and +2. If this range turns out not to contain the maximum of the log-likelihood, then the range can be expanded. To test H_0: $\lambda = 1$, calculate the likelihood-ratio test statistic

$$G_0^2 = -2 \times [\log_e L(\lambda = 1) - \log_e L(\lambda = \hat{\lambda})]$$

which is distributed as χ_1^2 under H_0. Alternatively, a 95% confidence interval for λ includes those values for which $\log_e L(\lambda) > \log_e L(\lambda = \hat{\lambda}) - 1/2 \times 1.96^2$, where $1.96^2 = \chi^2_{1,0.05}$.

Figure 9.2 shows a plot of the maximized log-likelihood against λ for Ornstein's interlocking-directorate regression. The maximum-likelihood estimate of λ is $\hat{\lambda} = 0.30$, and a 95% confidence interval, marked out by the intersection of the line near the top of the graph with the log-likelihood, runs from 0.20 to 0.41. (Recall that in Chapter 6 we employed a square-root transformation for these data to stabilize the error variance.)

Atkinson (1985) has proposed an approximate score test for the Box-Cox model, based on the constructed variable $G_i = y_i \times [\log_e(y_i/\tilde{y}) - 1]$, where $\tilde{y}$ is the geometric mean of y: $\tilde{y} = (y_1 \times y_2 \times \ldots \times y_n)^{1/n}$. This constructed variable is obtained by a linear approximation to the Box-Cox transformation $y^{(\lambda)}$ evaluated at $\lambda = 1$. The augmented model is then

$$y_i = \beta_0 + \beta_1 x_{1i} + \ldots + \beta_k x_{ki} + \varphi G_i + \varepsilon_i$$

The t test of H_0: $\varphi = 0$, that is, $t_0 = \hat{\varphi}/\mathrm{SE}(\hat{\varphi})$, assesses the need for a transformation. An estimate of λ (though not the MLE) is given by

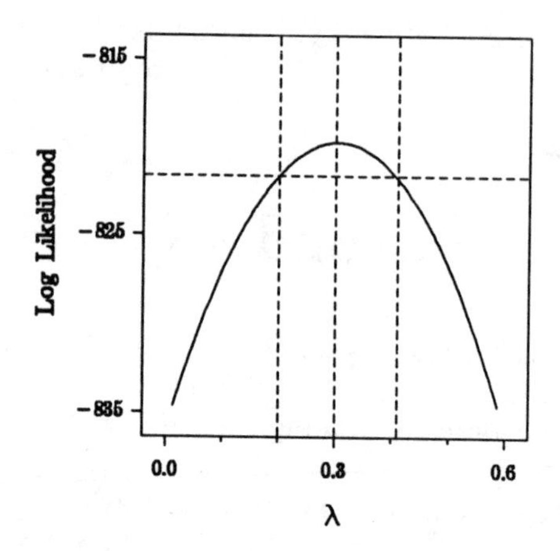

Figure 9.2. The conditional maximized log-likelihood in the Box-Cox model as a function of the transformation parameter λ, for Ornstein's interlocking-directorate regression. The intersection of the horizontal line near the top of the graph with the log-likelihood marks off a 95% confidence interval for λ.

$\widetilde{\lambda} = 1 - \hat{\varphi}$; and a partial-regression plot for G shows influence on $\hat{\varphi}$ and hence on the choice of λ.

Atkinson's constructed-variable plot for the interlocking-directorate regression is shown in Figure 9.3. Although the trend in the plot is not altogether linear, it appears that evidence for the transformation of y is spread through the data and does not depend unduly on a small number of observations. The coefficient of the constructed variable in the regression is $\hat{\varphi} = 0.588$, with $\mathrm{SE}(\hat{\varphi}) = 0.032$, providing very strong evidence of the need to transform y. The suggested transformation, $\widetilde{\lambda} = 1 - 0.588 = 0.412$, is close to the MLE.

Box-Tidwell Transformation of the *x*s

Now, consider the model

$$y_i = \beta_0 + \beta_1 x_{1i}^{\gamma_1} + \ldots + \beta_k x_{ki}^{\gamma_k} + \varepsilon_i$$

$$\varepsilon_i \sim \mathrm{NID}(0, \sigma^2)$$

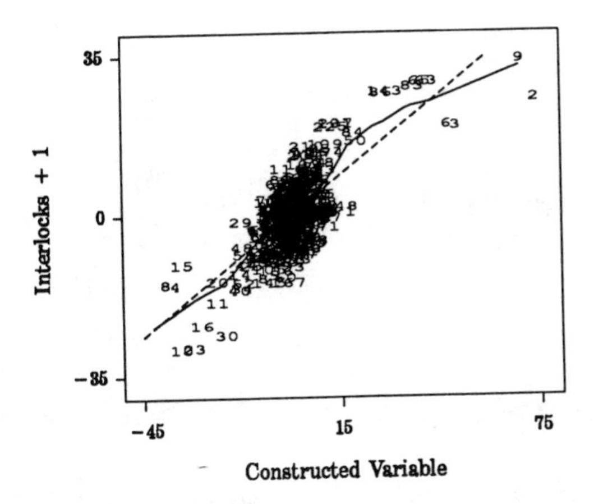

Figure 9.3. Constructed-variable plot for the Box-Cox transformation of Ornstein's interlocking-directorate regression. The observation index is plotted for each point. Both the least-squares (broken line) and lowess (solid line) regressions are shown.

It is assumed that all of the x_{ji} are positive. The parameters of this model, $\beta_0, \beta_1, \ldots, \beta_k$ and $\gamma_1, \ldots, \gamma_k$, could be estimated along with σ^2 by general nonlinear least-squares methods (see, e.g., Gallant, 1975), but Box and Tidwell (1962) suggest instead a more efficient procedure that also yields a constructed-variable diagnostic:

1. Regress y on $x_1, \ldots, x_k$, obtaining $b_0, b_1, \ldots, b_k$.
2. Regress y on $x_1, \ldots, x_k$ and the constructed variables $x_1 \log_e x_1, \ldots, x_k \log_e x_k$, obtaining $b_0', b_1' \ldots, b_k'$ and $d_1, \ldots, d_k$. Note that because of the presence of the constructed variables in this second regression, in general $b_j \neq b_j'$. As in the Box-Cox model, the constructed variables result from a linear approximation to x_j^γ evaluated at $\gamma = 1$.
3. The constructed variable $x_j \log_e x_j$ may be used to assess the need for a transformation of x_j by testing the hypothesis H_0: $\delta_j = 0$, where δ_j is the population coefficient of $x_j \log_e x_j$ in the second regression. Partial-regression plots for the constructed variables are useful for assessing leverage and influence on the decision to transform the xs.
4. An estimate of γ_j is given by $\tilde{\gamma}_j = 1 + d_j / b_j$. Recall that b_j is from the *initial* (step 1) regression.

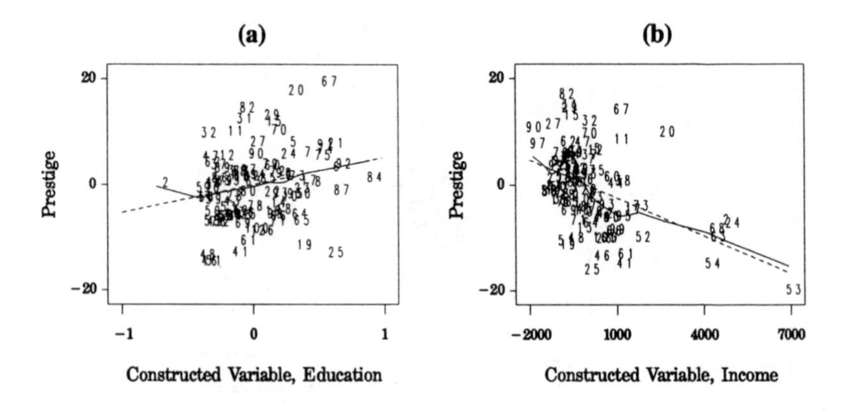

Figure 9.4. Constructed-variable plot for the Box-Tidwell transformation of (a) education and (b) income in the Canadian occupational-prestige regression. The observation index is plotted for each point. Both the least-squares (broken line) and lowess (solid line) regressions are shown.

This procedure can be iterated through steps 1, 2, and 4 until the estimates of the transformation parameters stabilize, yielding the MLEs $\hat{\gamma}_j$.

For the Canadian occupational-prestige data, leaving the regressors for percentage of women (W and W^2) untransformed, the coefficients of $E \log_e E$ and $I \log_e I$ in the auxiliary regression are, respectively, $d_E = 5.30$ with $\text{SE}(d_E) = 2.20$, and $d_I = -0.00243$ with $\text{SE}(d_I) = 0.00046$. There is, consequently, much stronger evidence of the need to transform income than education. Recall from Chapter 7 that the power transformation of education is not wholly appropriate. The first-step estimates of the transformation parameters are

$$\tilde{\gamma}_E = 1 + d_E / b_E = 1 + 5.30/4.26 = 2.2$$

$$\tilde{\gamma}_I = 1 + d_I / b_I = 1 - 0.00243/0.00127 = -0.91$$

The fully iterated MLEs of the transformation parameters are $\hat{\gamma}_E = 2.2$ and $\hat{\gamma}_I = -0.038$. Compare these values with the square and log transformations discovered following trial and error in Chapter 7. Constructed-variable plots for the transformation of education and income, in Figure 9.4, suggest that there is general evidence for these transformations, although there are some high-leverage observations in the income plot.

Nonconstant Error Variance Revisited

Breusch and Pagan (1979) develop a score test for heteroscedasticity based on the specification that the error variance depends on known variables $z_1, \ldots, z_p$, and may be modeled as

$$\sigma_i^2 = V(\varepsilon_i) = g(\gamma_0 + \gamma_1 z_{1i} + \ldots + \gamma_p z_{pi})$$

where the function $g(\bullet)$ is quite general. The same test was independently derived by Cook and Weisberg (1983). The score statistic for the hypothesis that the σ_i^2 are constant, which is equivalent to H_0: $\gamma_1 = \ldots = \gamma_p = 0$, may be formulated as an auxiliary regression problem.

Let $u_i = e_i^2/\hat{\sigma}^2$, where $\hat{\sigma}^2 = \Sigma\, e_i^2/n$ is the MLE of the error variance (note the division by n rather than degrees of freedom $n-k-1$). The u_i are a type of squared standardized residuals. Regress u on the zs:

$$u_i = \alpha_0 + \alpha_1 z_{1i} + \ldots + \alpha_p z_{pi} + \omega_i \qquad [9.1]$$

Breusch and Pagan show that the score statistic $S^2 = \Sigma(\hat{u}_i - \bar{u})^2/2$ is asymptotically distributed as χ_p^2 under H_0: $\sigma_i^2 = \sigma^2$. Here, the $\hat{u}_i$ are fitted values from the regression of u on the zs and thus S^2 is half the regression sum of squares from fitting Equation 9.1.

In applications, it is of course necessary to select zs, the choice of which depends on the suspected pattern of nonconstant error variance. If several patterns are suspected, then several score tests may be performed. Employing $x_1, \ldots, x_k$ in the auxiliary regression Equation 9.1, for example, permits detection of a tendency of the error variance to increase with values of one or more of the independent variables in the main regression.

Likewise, Cook and Weisberg (1983) suggest regressing u on the fitted values from the main regression (i.e., fitting the auxiliary regression $u_i = \alpha_0 + \alpha_1\hat{y}_i + \omega_i$), producing a one-degree-of-freedom score test to detect the common tendency of the error variance to increase with the level of the dependent variable. When the error variances in fact follow this pattern, the auxiliary regression of u on $\hat{y}$ provides a more powerful test than the more general regression of u

on the xs. A similar (but more complex) procedure is described by Anscombe (1961), who suggests correcting detected heteroscedasticity by transforming y to $y^{(\tilde{\lambda})}$, with $\tilde{\lambda} = 1 - 1/2\hat{\alpha}_1\bar{y}$.

Finally, White (1980) proposes a similar score test based on a comparison of his heteroscedasticity-corrected estimator of coefficient sampling variance (see Chapter 6 and Appendix A6.3) with the usual estimator of coefficient variance. If the two estimators are sufficiently different, then doubt is cast on the assumption of constant error variance. White's test may be implemented as an auxiliary regression of the squared residuals from the main regression, e_i^2, on all of the xs together with all squares and pairwise products of the xs. Thus, for k = 2 independent variables in the main regression, we would fit the model

$$e_i^2 = \delta_0 + \delta_1 x_{1i} + \delta_2 x_{2i} + \delta_3 x_{1i}^2 + \delta_4 x_{2i}^2 + \delta_5 x_{1i} x_{2i} + \nu_i$$

In general, there will be $p = k(k + 3)/2$ terms in the auxiliary regression, plus the constant.

The score statistic for testing the null hypothesis of constant error variance is $S^2 = nR^2$, where R^2 is the squared multiple correlation from the auxiliary regression. Under the null hypothesis, S^2 follows an asymptotic χ^2 distribution with p degrees of freedom.

Because all of these score tests are potentially sensitive to violations of model assumptions other than constant error variance, it is important in practice to supplement the tests with graphical diagnostics, as suggested by Cook and Weisberg (1983). When there are several zs, a simple diagnostic is to plot u_i against $\hat{u}_i$, the fitted value from the auxiliary regression. We also can construct partial-regression plots for the zs in the auxiliary regression. When u_i is simply regressed on $\hat{y}_i$, these plots are essentially similar to the plot of studentized residuals against fitted values proposed in Chapter 6.

Applied to Ornstein's interlocking-directorate data, an auxiliary regression of u on $\hat{y}$ yields $\hat{u} = 0.134 + 0.0594\,\hat{y}$, and $S^2 = 147.6/2 = 73.8$ on one degree of freedom. There is, consequently, very strong evidence for nonconstant error variance. The suggested variance-stabilizing transformation using Anscombe's rule is $\tilde{\lambda} = 1 - 1/2(0.0594)(14.81) = 0.56$. Compare this value with those produced by the Box-Cox model ($\hat{\lambda} = 0.3$) and by trial and error ($\lambda = 0.5$, from Chapter 6).

An auxiliary regression of u on the independent variables in the main regression yields $S^2 = 172.6/2 = 86.3$ on $k = 13$ degrees of freedom, and thus also provides strong evidence against constant error variance. Examination of the coefficients from the auxiliary regression (not shown) indicates in particular a tendency of the error variance to increase with assets. Note, however, that the score statistic for the more general test is not much larger than for the regression of u on $\hat{y}$, suggesting that the pattern of nonconstant error variance is indeed for the spread of the errors to increase with the level of y. Assets are, of course, an important component of $\hat{y}$. Because White's test requires 104 regressors for this problem, it was not performed.

10. RECOMMENDATIONS

1. Screen your data prior to undertaking a complex statistical analysis. Examining univariate distributions and bivariate plots, although not substitutes for the methods presented in this monograph, often will reveal problems such as wild data values, highly skewed distributions, extreme nonlinearity, and so on. If the data set is small, consider entering the data into the computer yourself. Generally, do not hesitate to dirty your hands with the data.

2. Employ a small number of simple, robust, and informative diagnostics as a matter of course, following up problems that are revealed with more specialized and sophisticated methods. The following selection of everyday diagnostics is a reasonable choice:

a. *Collinearity:* Although collinearity is rarely a serious problem for individual-level cross-sectional data—more frequently with aggregated or longitudinal data—it is simple and informative to compute variance-inflation factors.

b. *Influential data, outliers, and non-normality:* Except in the case of grossly inaccurate data (e.g., missing data codes treated as valid data), influential data are much more common in small data sets than in large ones. A plot of studentized residuals by hat-values is a good diagnostic, because all of the influence statistics that I presented depend in one way or another on these quantities or on closely related values. An index plot of Cook's D (or of DFFITS) provides a summary measure of influence on the regression coefficients. Partial-regression plots are useful for displaying leverage and influence on individual coefficients and may reveal influential subsets of observations not apparent from single-observation deletion statistics like Cook's D. A normal quantile-comparison plot of studentized residuals reveals outliers as well as skewed and heavy-tailed error distributions. A stem-and-leaf display, histogram, or smoothed histogram of studentized residuals conveys a good impression of the shape of the residual distribution and may reveal problems such as multiple modes.

c. *Nonlinearity:* If there appears to be nonlinearity in a partial-regression plot, the pattern may be revealed more clearly in a partial-residual plot. Because the latter is very easily constructed, it can be used routinely.

d. *Nonconstant error variance:* A plot of studentized residuals (or their squares or absolute values) against fitted values will reveal a tendency of the error variance to change with the level of y, which is the most common problem of this type.

3. If possible, employ smoothing methods, like lowess smoothing of scatterplots, to help reveal patterns in the data. Although diagnostic techniques are most crucially directed towards gross rather than subtle problems, it is at times important to compensate for the tendency to perceive nonexistent visual patterns and for the difficulty in other contexts of separating information from visual noise.

4. Try to avoid the pitfalls of "over-fitting" (i.e., modifying a statistical model to capture chance features of the data). Part of the art of data analysis is judging how far to accommodate data. At one extreme, there are those who ignore unanticipated patterns in the data so as to implement a textbook caricature of "objective" statistical analysis that requires the model to be specified completely in advance. At the other extreme, there are those—often recent students of diagnostic techniques—who trim away large portions of their data, or who transform data so as to achieve a trivially "better" fit.

Discarding discrepant data is satirized in Figure 10.1. At the risk of ruining a good joke, I feel compelled to point out that it would be worse to fit a line to all of the data in this figure than to eliminate the offending points. It would, of course, be best to try to understand what differentiates the discrepant cluster of points from the rest of the data.

Although the formal statistical analysis of "specification searches"—that is, choice of statistical model informed partly by examination of the data—is exceedingly complex (see Leamer, 1978), it

Figure 10.1. Regression diagnostics in action.

SOURCE: Reprinted with permission from the announcement of the Summer Program of the Inter-University Consortium for Political and Social Research, 1990.

is often possible to assess honestly the adequacy of a model by cross-validation (Mosteller and Tukey, 1977). To cross-validate results, we first divide our sample randomly into two parts, possibly but not necessarily of equal size. The first subsample is employed to choose a statistical model for the data; the model is then validated on the second subsample. This approach is particularly useful when the data have been employed to select a subset of independent variables in a regression, or when transformations have been used to deal with problems such as nonlinearity. Cross-validation does not speak as clearly to problems of outlying and influential data: These problems pertain to individual observations which are, of course, not shared by the two subsamples.

Researchers are reluctant to divide their samples, as required by cross-validation, because they are sensitive to the impact of sample size on precision of estimation and the power of statistical tests. Yet, when the data are used partly to choose the model and partly to estimate its parameters, the gain in precision is illusory, at least to a degree. What is worst in my opinion, however, is to avoid examining the adequacy of an initially specified model simply to protect the "purity" of classical estimates and tests.

As mentioned, the elimination of outlying and influential data does not lend itself to check by cross-validation. Yet, estimates of sampling variation following these procedures are probably optimistic. If the rules employed for rejecting outliers or influential data can be stated precisely, however, it is possible to estimate sampling variation empirically (see, e.g., Diaconis and Efron, 1983; Stine, 1990).

5. Take account of the sampling properties of the data. Data that arise from complex sampling designs generally have non-negligible dependencies among the observations (e.g., Kish, 1965). Likewise, substantial quantities of missing data require special treatment (e.g., Little and Rubin, 1990).

A common situation in which the assumption of independent errors usually is not reasonable occurs when the observations are defined by points in time, producing what are commonly termed *time-series* data. Methods for detecting and dealing with autocorrelated errors in time-series regression may be found, for example, in Ostrom (1990) and Kmenta (1986, Ch. 8). A useful and familiar preliminary diagnostic in this situation is to plot least-squares residuals against the observation index, which represents time.

Computing Diagnostics

Standard statistical software packages (such as SAS, SPSS, BMDP, and Systat) now directly incorporate many of the diagnostics discussed in this monograph. Even when a particular statistic or procedure is not offered directly by a statistical package, it is often simple to compute or construct. DFFITS and COVRATIO, for example, which are discussed in Chapter 4, may be computed from hat-values and studentized residuals. Likewise, the partial-regression and constructed-variable plots discussed in Chapters 4 and 9 may be obtained by plotting residuals from appropriately formulated regressions. Even relatively complex procedures such as lowess can be programmed in the macro languages provided by packages such as SAS (and indeed scatterplot smoothers like lowess are already included in Systat and some other packages).

Almost all of the computing for this monograph was done with the PC version of SAS, and the graphs that appear were (with a couple of exceptions) constructed with SAS/GRAPH. Current statistical software therefore is no obstacle to the application of diagnostic techniques, and it is likely that the diagnostic capabilities of standard programs will continue to improve.

Further Reading

There is voluminous journal literature on regression diagnostics and related topics, such as exploratory and graphical data analysis. Fortunately, there are now several texts that present this literature in a more digestible form.

In my opinion, Cook and Weisberg's (1982b) text is the best book-length presentation of methods for assessing leverage (a term that the authors dislike), outliers, and influence. The book also includes good discussions of other problems, such as nonlinearity and transformations of the dependent and independent variables, but there is no treatment of collinearity. Cook and Weisberg's (1982a) article presents in condensed form many of the topics that appear in their book.

Chatterjee and Hadi (1988) is a thorough and up-to-date text that deals primarily with influential data but touches on other topics such as nonlinearity and nonconstant error variance. The book is distinguished by comparative considerations of alternative measures of influence on a variety of regression outputs, including regression coefficients, coefficient variances, and collinearity.

Atkinson (1985) also is a valuable source, emphasizing the author's important contributions to regression diagnostics, such as constructed-variable plots and simulation methods. Likewise, Belsley et al. (1980) is a seminally important book dealing with influential data and collinearity, primarily presenting the authors' work in these areas. I believe, however, that their treatment of collinearity is marred by an argument that the intercept should not be "adjusted out" prior to assessing ill-conditioning (see the discussion following Belsley, 1984).

Several applied-regression and linear-model texts have relatively strong treatments of diagnostics. See, in particular, Chatterjee and Price (1977), Daniel and Wood (1980), and Draper and Smith (1981) for discussions of collinearity, variable selection, and some information on residual-based diagnostics; and Weisberg (1985) and Fox (1984) for treatments of a variety of topics covered in this monograph. Likewise, general econometrics books often contain information on detecting and correcting violations of regression-model assumptions, generally with a more theoretical than data-analytic flavor. See Kmenta (1986) for a strong example of this approach.

There are many good books on graphical and exploratory methods of data analysis, including introductory texts on graphics by Cleveland (1985) and on exploratory data analysis by Velleman and Hoaglin (1981). Also see Chambers et al. (1983), which has contents similar to Cleveland's book; Tukey's (1977) original—and iconoclastic—presentation of exploratory data analysis; and books of readings edited by Hoaglin, Mosteller, and Tukey (1983, 1985) and by Fox and Long (1990). Finally, Mosteller and Tukey's (1977) idiosyncratic regression text, a sort of companion volume to Tukey (1977), contains a variety of interesting material on regression from a strongly data-analytic perspective.

APPENDIX

A2.1 Least-Squares Fit, Joint Confidence Regions, and Tests

In matrix form, the linear-regression model is written $\mathbf{y} = \mathbf{X}\beta + \varepsilon$ where $\mathbf{y}$ is an $n \times 1$ vector of dependent-variable values; $\mathbf{X}$ is an $n \times (k + 1)$ matrix of regressors, including the constant regressor of 1s as its first column; β is a $(k + 1) \times 1$ vector of regression parameters; and ε is an $n \times 1$ vector of errors. By assumption, $\varepsilon \sim N_n(0, \sigma^2 \mathbf{I})$, independent of $\mathbf{X}$.

The fitted model is $\mathbf{y} = \mathbf{Xb} + \mathbf{e}$. To obtain the least-squares estimates $\mathbf{b}$ of β, we need to minimize the sum of squared residuals, $\mathbf{e}'\mathbf{e} = [\text{length}(\mathbf{e})]^2$. Because $\mathbf{e} = \mathbf{y} - \hat{\mathbf{y}}$, the length of $\mathbf{e}$ is minimized by taking $\hat{\mathbf{y}} = \mathbf{Xb}$ as the orthogonal projection of $\mathbf{y}$ onto the subspace spanned by the columns of $\mathbf{X}$. Then, because $\mathbf{X}'\mathbf{e} = \mathbf{0}$, we have $\mathbf{X}'\mathbf{Xb} = \mathbf{X}'\mathbf{y}$, which are the normal equations in matrix form. Note that because $\hat{\mathbf{y}}$ is in the column subspace of $\mathbf{X}$, the residuals and fitted values are orthogonal: $\Sigma e_i \hat{y}_i = \mathbf{e}'\hat{\mathbf{y}} = 0$. Moreover, because the first column of $\mathbf{X}$ consists of ones, $\Sigma e_i = \mathbf{1}'\mathbf{e} = 0$.

Alternatively, but equivalently,

$$\mathbf{e}'\mathbf{e} = (\mathbf{y} - \mathbf{Xb})' (\mathbf{y} - \mathbf{Xb})$$

$$= \mathbf{y}'\mathbf{y} - 2\mathbf{y}' \mathbf{Xb} + \mathbf{b}' \mathbf{X}' \mathbf{Xb}$$

Differentiating produces $\partial \mathbf{e}'\mathbf{e}/\partial \mathbf{b} = -2\mathbf{X}'\mathbf{y} + 2\mathbf{X}'\mathbf{Xb}$, and setting the partial derivatives to zero so as to minimize the sum-of-squares function gives the normal equations. If $\mathbf{X}'\mathbf{X}$ is nonsingular, which will occur as long as the columns of $\mathbf{X}$ are not collinear, then $\mathbf{b} = (\mathbf{X}'\mathbf{X})^{-1}\mathbf{X}'\mathbf{y}$.

Because, by assumption, $E(\varepsilon) = \mathbf{0}$, it follows that $E(\mathbf{y}) = \mathbf{X}\beta$, and $E(\mathbf{b}) = (\mathbf{X}'\mathbf{X})^{-1}\mathbf{X}'E(\mathbf{y}) = \beta$; thus $\mathbf{b}$ is an unbiased estimator of β. Likewise, by assumption, $V(\mathbf{y}) = V(\varepsilon) = \sigma^2\mathbf{I}$, and hence, using the symmetry of the sum-of-squares-and-products matrix $\mathbf{X}'\mathbf{X}$,

$$V(\mathbf{b}) = (\mathbf{X}'\mathbf{X})^{-1} \mathbf{X}' V(\mathbf{y}) [(\mathbf{X}'\mathbf{X})^{-1} \mathbf{X}']' = \sigma^2 (\mathbf{X}'\mathbf{X})^{-1}$$

Under the assumption of normally distributed errors, then,

$$\mathbf{b} \sim N_{k+1} [\beta, \sigma^2(\mathbf{X}'\mathbf{X})^{-1}].$$

A 100(1–α)% ellipsoidal joint confidence region for the regression coefficients is given by

$$(\mathbf{b} - \beta)' (\mathbf{X}'\mathbf{X}) (\mathbf{b} - \beta) \leq (k+1)\, s^2 F_{\alpha, k+1, n-k-1}$$

where $s^2 = \mathbf{e}'\mathbf{e}/(n - k - 1)$ is an unbiased estimator of σ^2, and $F_{\alpha,k+1,n-k-1}$ is the α critical value of F with $k + 1$ and $n - k - 1$ degrees of freedom. For a subset β_1 of p regression coefficients, we have the 100(1 – α)% confidence region

$$(\mathbf{b}_1 - \beta_1)' \, \mathbf{V}_{11}^{-1} \, (\mathbf{b}_1 - \beta_1) \leq ps^2 F_{\alpha, p, n-k-1} \qquad \text{[A.1]}$$

Here $\mathbf{V}_{11}$ is the $p \times p$ submatrix of rows and columns of $(\mathbf{X}'\mathbf{X})^{-1}$ corresponding to the entries of $\mathbf{b}_1$.

F tests are easily obtained from the expressions for these confidence regions. For example, to test H_0: $\beta_1 = \beta_1^{(0)}$, find

$$F_0 = \frac{(\mathbf{b}_1 - \beta_1^{(0)})' \, \mathbf{V}_{11}^{-1} \, (\mathbf{b}_1 - \beta_1^{(0)})}{ps^2}$$

which is distributed as $F_{p,n-k-1}$ under H_0. For $\beta_1^{(0)} = \mathbf{0}$, F_0 is simply the incremental F statistic given in the text.

A3.1 Ridge Regression

Ridge regression (Hoerl and Kennard, 1970a, 1970b) is an attempt to obtain more efficient estimates in the presence of strong collinearity. My primary objective in explaining ridge regression here is to suggest that it is not a general remedy for collinearity.

Begin by rescaling $\mathbf{y}$ and the columns of $\mathbf{X}$ to zero means and unit length, so that sums of products are correlations. Then the ridge estimator of the standardized regression coefficients is

$$\mathbf{b}_z^* = (\mathbf{R}_{xx} + z\mathbf{I})^{-1} \, \mathbf{r}_{xy} = (\mathbf{R}_{xx} + z\mathbf{I})^{-1} \, \mathbf{R}_{xx} \, \mathbf{b}^*$$

where $\mathbf{b}^* = \mathbf{R}_{xx}^{-1}\mathbf{r}_{xy}$ is the least-squares estimator, and $z \geq 0$ is the *ridge constant,* which is selected by the researcher. Here, $\mathbf{R}_{xx}$ is the matrix of correlations among the xs, and $\mathbf{r}_{xy}$ is the vector of correlations between the xs and y. Intuitively, by adding z to each diagonal entry of $\mathbf{R}_{xx}$, the diagonal entries (originally 1) are inflated relative to the off-diagonal entries (the correlations among the regressors), thus improving the "conditioning" of the independent-variable correlation matrix. When $z = 0$, the least-squares and ridge estimators coincide: $\mathbf{b}_0^* = \mathbf{b}^*$.

Hoerl and Kennard show that the bias of $\mathbf{b}_z^*$ increases with z; that for $z > 0$, $V(\mathbf{b}_z^*) < V(\mathbf{b}^*)$; that $V(\mathbf{b}_z^*)$ declines as z increases; and that there is always a range of values of z for which $\text{MSE}(\mathbf{b}_z^*) < \text{MSE}(\mathbf{b}^*)$. Recall that mean-squared error is the sum of sampling variance and squared bias; the trick in ridge regression, therefore, is to pick z so that the trade-off of bias against variance is favorable.

Values of z for which the ridge estimator is better than the least-squares estimator depend on the unknown parameters β^*; it is therefore not obvious how the theoretical advantage of the ridge estimator can be realized in practice. Draper and Smith (1981) demonstrate that the choice of z for the ridge estimator implicitly places a constraint on the length of the estimated parameter vector $\mathbf{b}_z^*$.

A4.1 Hat-Values and the Hat Matrix

The fitted values in least-squares regression are a linear function of the observed ys:

$$\hat{\mathbf{y}} = \mathbf{Xb} = \mathbf{X}(\mathbf{X}'\mathbf{X})^{-1}\mathbf{X}'\mathbf{y} = \mathbf{Hy}$$

Here $\mathbf{H} = \mathbf{X}(\mathbf{X}'\mathbf{X})^{-1}\mathbf{X}'$ is the "hat matrix," so named because it transforms $\mathbf{y}$ into $\hat{\mathbf{y}}$. The hat matrix is symmetric ($\mathbf{H} = \mathbf{H}'$) and idempotent ($\mathbf{H}^2 = \mathbf{H}$), as can easily be verified. Consequently, the diagonal entries of the hat matrix $h_i = h_{ii}$, called *hat-values*, are

$$h_i = \sum_{j=1}^{n} h_{ij}^2 = h_i^2 + \sum_{j \neq i} h_{ij}^2$$

which implies that $0 \leq h_{ii} \leq 1$. If $\mathbf{X}$ includes the constant regressor, then $1/n \leq h_i$. Finally, because $\mathbf{H}$ is a projection matrix, projecting $\mathbf{y}$ orthogonally onto the subspace spanned by the columns of $\mathbf{X}$, it follows that $\Sigma h_i = k + 1$, and thus $\bar{h} = (k + 1)/n$, as stated in the text. See Hoaglin and Welsch (1978) or Chatterjee and Hadi (1988, Ch. 2) for details.

A4.2 The Distribution of the Least-Squares Residuals

The least-squares residuals are given by

$$\mathbf{e} = \mathbf{y} - \hat{\mathbf{y}} = (\mathbf{X}\boldsymbol{\beta} + \boldsymbol{\varepsilon}) - \mathbf{X}(\mathbf{X}'\mathbf{X})^{-1}\mathbf{X}'(\mathbf{X}\boldsymbol{\beta} + \boldsymbol{\varepsilon}) = (\mathbf{I} - \mathbf{H})\boldsymbol{\varepsilon}$$

Thus,

$$E(\mathbf{e}) = (\mathbf{I} - \mathbf{H})E(\boldsymbol{\varepsilon}) = (\mathbf{I} - \mathbf{H})\mathbf{0} = \mathbf{0}$$

and

$$V(\mathbf{e}) = (\mathbf{I} - \mathbf{H})\, V(\mathbf{e})\, (\mathbf{I} - \mathbf{H})' = \sigma^2 (\mathbf{I} - \mathbf{H})$$

because $\mathbf{I} - \mathbf{H}$, like $\mathbf{H}$, is symmetric and idempotent. The matrix $\mathbf{I} - \mathbf{H}$ is not diagonal, and its diagonal entries are usually unequal; the residuals, therefore, are correlated and usually have unequal variances, even though the errors are, by assumption, independent with equal variances.

A4.3 Deletion Diagnostics

Let $\mathbf{b}_{(-i)}$ denote the vector of least-squares regression coefficients calculated with the ith observation omitted. Then $\mathbf{d}_i = \mathbf{b} - \mathbf{b}_{(-i)}$ represents the influence of observation i on the regression coefficients; $\mathbf{d}_i$ may calculated efficiently by

$$\mathbf{d}_i = (\mathbf{X}'\mathbf{X})^{-1}\, \mathbf{x}_i \frac{e_i}{1 - h_i} \qquad [A.2]$$

where $\mathbf{x}_i'$ is the ith row of $\mathbf{X}$.

Cook's D_i is the F value for testing the "hypothesis" that $\beta = \mathbf{b}_{(-i)}$:

$$D_i = \frac{(\mathbf{b} - \mathbf{b}_{(-i)})'\, \mathbf{X}'\mathbf{X}\, (\mathbf{b} - \mathbf{b}_{(-i)})}{(k+1)\, s^2}$$

$$= \frac{(\hat{\mathbf{y}} - \hat{\mathbf{y}}_{(-i)})'\, (\hat{\mathbf{y}} - \hat{\mathbf{y}}_{(-i)})}{(k+1)\, s^2}$$

An alternative interpretation of D, therefore, is that it measures the aggregate influence of observation i on the fitted values $\hat{\mathbf{y}}$, which is why Belsley et al. (1980) call their similar statistic "DFFITS." Using Equation A.2,

$$D_i = \frac{e_i^2}{s^2 (k+1)} \times \frac{h_i}{(1 - h_i)^2} = \frac{e_i'^2}{k+1} \times \frac{h_i}{1 - h_i}$$

which is the formula given in the text.

A4.4 The Partial-Regression Plot

In vector form, the fitted multiple-regression model is

$$\mathbf{y} = b_0\mathbf{1} + b_1\mathbf{x}_1 + \ldots + b_k\mathbf{x}_k + \mathbf{e} \qquad [A.3]$$

where $\mathbf{y}$ and the $\mathbf{x}_j$ are $n \times 1$ vectors of observations, $\mathbf{1}$ is an $n \times 1$ vector of ones, and $\mathbf{e}$ is the $n \times 1$ residual vector. In least-squares regression, $\hat{\mathbf{y}} = b_0\mathbf{1} + b_1\mathbf{x}_1 + \ldots + b_k\mathbf{x}_k$ is the orthogonal projection of $\mathbf{y}$ onto the subspace spanned by the regressors. Let $\mathbf{y}^{(1)}$ and $\mathbf{x}^{(1)}$ be the projections of $\mathbf{y}$ and $\mathbf{x}_1$, respectively, onto the orthogonal complement of the subspace spanned by $\mathbf{1}$ and $\mathbf{x}_2, \ldots, \mathbf{x}_k$ (i.e., the residual vectors from the regressions of y and x_1 on the other xs). Then, by the geometry of projections, the orthogonal projection of $\mathbf{y}^{(1)}$ onto $\mathbf{x}^{(1)}$ is $b_1\mathbf{x}^{(1)}$, and $\mathbf{y}^{(1)} - b_1\mathbf{x}^{(1)} = \mathbf{e}$, the residual vector from the overall regression in Equation A.3.

A6.1 Smoothing Scatterplots by Lowess

An acronym for *lo*cally *we*ighted *s*catterplot *s*moother, *lowess* (Cleveland, 1985) produces a smoothed fitted value $\hat{y}_i$ corresponding to each x_i (where y and x are used generically for the vertical and horizontal variables in the scatterplot). To find the smoothed values, the lowess procedure fits n regression lines to the data, one for each observation i, emphasizing the points with x values near x_i. The lowess procedure is illustrated in Figure A.1. Lowess is computationally intensive and thus requires a special computer program for implementation, but such programs are simple to write and are increasingly common.

1. *Choose a smoothing fraction:* Select a fraction of the data $0 < f \leq 1$ to include in each fit, corresponding to $r = [fn]$ data values, where the square brackets denote rounding to the nearest integer. Often $f = 1/2$ or 2/3 works well. Larger values of f produce smoother results.

2. *Locally weighted regressions:* For each x_i, select the r values of x closest to it, denoted $x_1^{(i)}, \ldots, x_r^{(i)}$ (see Figure A.1a). The window half-width for this observation is then the distance to the farthest $x_j^{(i)}$: that is, $W_i = |x_i - x_r^{(i)}|$. For each of the r observations in the window, compute the weight $w_j^{(i)} = w_t[(x_j^{(i)} - x_i) / W_i]$, where w_t is the tricube weight function

$$w_t(z) = \begin{cases} 0 & \text{for } |z| \geq 1 \\ (1 - |z|^3)^3 & \text{for } |z| < 1 \end{cases}$$

(Here, z simply represents the argument of the tricube function—i.e., $(x_j^{(i)} - x_i)/W_i$.) Thus $w_j^{(i)}$ descends to zero as $x_j^{(i)}$ approaches the

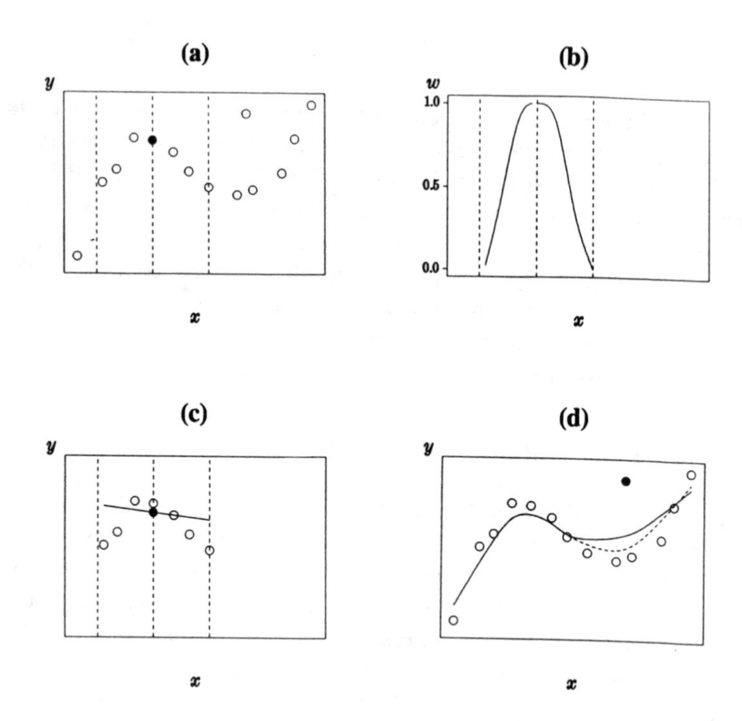

Figure A.1. How lowess works: Panels a, b, and c show the calculation of the lowess fit $\hat{y}_5$ at $x = x_5$ for a dataset of $n = 14$ observations. (a) A window is located with center at x_5 to include $f = ½$ of the data; thus $r = [fn] = 7$ points are within the window. The point (x_5, y_5) is shown as a filled dot. (b) The tricube weight function descends to zero at the boundaries of the window and attains its maximum value of one at $x = x_5$. (c) A local regression line is fit to the seven observations in the window, employing weights given by the weight function in (b). The lowess fitted value $\hat{y}_5$ at x_5 is shown as a filled dot. Steps a, b, and c are repeated for each observation to obtain the full set of 14 fitted values. (d) The lowess curve (solid line) is obtained by connecting the fitted values $\hat{y}_1, \ldots, \hat{y}_{14}$. Note that the curve is pulled toward the outlying observation (filled dot). The dotted line shows how downweighting the outlier produces a more robust fit (connecting the fitted values $\hat{y}'_1, \ldots, \hat{y}'_{14}$).

SOURCE: Adapted from *The Elements of Graphing Data*, by W. S. Cleveland. Copyright © 1985 Bell Telephone Laboratories, Inc., Murray Hill, NJ. Adapted by permission of Wadsworth & Brooks/Cole Advanced Books & Software, Pacific Grove, CA 93950.

boundaries of the window and is largest at x_i. (See Figure A.1b). Then fit the regression equation

$$y_j^{(i)} = a_i + b_i x_j^{(i)} + e_j^{(i)}$$

to minimize $\Sigma_{j=1}^{r} w_j^{(i)} e_j^{(i)2}$ (cf., Appendix A6.2 on weighted-least-squares regression). Compute the fitted value $\hat{y}_i = a_i + b_i x_i$ (see Figure A.1c). Note that one regression equation is fit, and one fitted value is calculated, for each $i = 1, \ldots, n$.

3. *Down-weight outliers:* Calculate residuals $e_i = y_i - \hat{y}_i$. Compute robustness weights that discount observations with large residuals: $\delta_i = w_b(e_i/6M)$, where M is the median of the absolute residuals $|e_i|$, and w_b is the bisquare weight function:

$$w_b(z) = \begin{cases} 0 & \text{for } |z| \geq 1 \\ (1 - z^2)^2 & \text{for } |z| < 1 \end{cases}$$

4. *Robust locally weighted regressions:* Repeat step 2, but use compound weights $\delta_j w_j^{(i)}$ in the individual regressions, finding new fitted values $\hat{y}_i'$ (see Figure A.1d). Steps 3 and 4 may be iterated to obtain more robustness.

A6.2 Weighted-Least-Squares Estimation

Suppose that in the regression model

$$y_i = \beta_0 + \beta_1 x_{1i} + \beta_2 x_{2i} + \ldots + \beta_k x_{ki} + \varepsilon_i \qquad \text{[A.4]}$$

$$\varepsilon_i \sim \text{NID}(0, \sigma_i^2)$$

the standard deviation of the errors is proportional to x_1, $\sigma_i = \sigma x_{1i}$. Dividing both sides of Equation A.4 by x_{1i} produces

$$\frac{y_i}{x_{1i}} = \beta_0 \frac{1}{x_{1i}} + \beta_1 + \beta_2 \frac{x_{2i}}{x_{1i}} + \ldots + \beta_k \frac{x_{ki}}{x_{1i}} + \frac{\varepsilon_i}{x_{1i}} \qquad \text{[A.5]}$$

and because $x_{1i} = \sigma_i/\sigma$, the last term becomes $\varepsilon_i' = \sigma\varepsilon_i/\sigma_i$. Because $V(\varepsilon_i') = \sigma^2 V(\varepsilon_i)/\sigma_i^2 = \sigma^2$ is constant, the transformed model in Equation A.5 may legitimately be fit by least-squares regression of y_i/x_{1i} on $1/x_{1i}$, a constant regressor, and x_{2i}/x_{1i} through x_{ki}/x_{1i}, producing estimates of the βs and their standard errors. The procedure is equivalent to minimizing the weighted sum of squares $\Sigma e_i^2/\sigma_i^2$, which yields maximum-likelihood estimates for Equation A.4. This general approach works whenever the error variances are known up to a constant of proportionality: $V(\varepsilon_i) = \sigma^2 w_i$. (See, for example, Weisberg [1985, Ch. 4].)

A6.3 Correcting Least-Squares Standard Errors for Heteroscedasticity

Recall from Appendix A2.1 that the covariance matrix of the least-squares estimator **b** is

$$V(\mathbf{b}) = (\mathbf{X}'\mathbf{X})^{-1}\mathbf{X}'V(\mathbf{y})\mathbf{X}(\mathbf{X}'\mathbf{X})^{-1} \quad \text{[A.6]}$$

Under the assumption of constant error variance, $V(\mathbf{y}) = \sigma^2\mathbf{I}$, and Equation A.6 simplifies to the usual formula, $V(\mathbf{b}) = \sigma^2(\mathbf{X}'\mathbf{X})^{-1}$. If, alternatively, the errors are heteroscedastic but independent, then $V(\mathbf{y}) = \Sigma = \text{diag}(\sigma_1^2, \ldots, \sigma_n^2)$, and

$$V(\mathbf{b}) = (\mathbf{X}'\mathbf{X})^{-1}\mathbf{X}'\Sigma\mathbf{X}(\mathbf{X}'\mathbf{X})^{-1}$$

Because $E(\varepsilon_i) = 0$, the variance of the ith error is $\sigma_i^2 = E(\varepsilon_i^2)$, which suggests the possibility of estimating $V(\mathbf{b})$ by

$$\widetilde{V}(\mathbf{b}) = (\mathbf{X}'\mathbf{X})^{-1}\mathbf{X}'\hat{\Sigma}\mathbf{X}(\mathbf{X}'\mathbf{X})^{-1} \quad \text{[A.7]}$$

with $\hat{\Sigma} = \text{diag}(e_1^2, \ldots, e_n^2)$, where e_i is the least-squares residual for observation i. White (1980) shows that Equation A.7 is a consistent estimator of $V(\mathbf{b})$.

For example, applied to Ornstein's interlocking-directorate data, White's approach produces estimated standard errors that are mostly similar to the estimates calculated by the usual formula (and given in Table 6.1): In fact, for most coefficients, the corrected standard errors are a bit smaller than the uncorrected ones. For the coefficient of square-root assets, however, the corrected standard error, 0.028, is substantially larger than the uncorrected one, 0.019.

A6.4 The Efficiency and Validity of Least-Squares Estimation When Error Variances Are Not Constant

The impact of nonconstant error variance on the efficiency of the ordinary least-squares (OLS) estimator and on the validity of least-squares inference depends on several factors, including the sample size, the degree of variation in σ_i^2, the configuration of x values, and the relationship between the error variance and the xs. It is therefore not possible to develop wholly general conclusions, but the following simple case is instructive and supports the advice given in the text.

Suppose that $y_i = \beta_0 + \beta_1 x_i + \varepsilon_i$, with $\varepsilon_i \sim \text{NID}(0, \sigma_i^2)$, and $\sigma_i = \sigma x_i$ (as in Appendix A6.2). Then the OLS estimator b_1 is less efficient than the WLS estimator $\hat{\beta}_1$, which under these circumstances is the maximally efficient unbiased estimator of β_1.

Formulas for the sampling variances of b_1 and $\hat{\beta}_1$ are easily derived (e.g., Kmenta, 1986, Ch. 8). The efficiency of the OLS estimator relative to the optimal WLS is given by $V(\hat{\beta}_1)/V(b_1)$, and the relative precision of OLS is the square root of this ratio, that is, $\text{SE}(\hat{\beta}_1)/\text{SE}(b_1)$.

Now, suppose that x is uniformly distributed over the interval $[x_0, \alpha x_0]$, where $x_0 > 0$ and $\alpha > 0$, so that α is the ratio of largest to smallest x (and, consequently, of largest to smallest σ_i). The relative precision of the OLS estimator stabilizes quickly as the sample size grows, and exceeds 90% when $\alpha = 2$, and 85% when $\alpha = 3$, even when n is as small as 20. For $\alpha = 10$, the penalty for using OLS is greater, but even here the relative precision of OLS exceeds 65% for $n \geq 20$.

The validity of statistical inferences based on least-squares estimation is even less sensitive to common patterns of nonconstant error variance. Here, we need to compare the expectation of the usual estimator of $V(b_1)$, which will typically be biased when error variance is not constant, with the true sampling variance of b_1. Again, the formula for $E[\hat{V}(b_1)]$ is simple to derive (and may be found in Kmenta, 1986, Ch. 8). The square root of $E[\hat{V}(b_1)]/V(b_1)$ expresses the result in relative standard-error terms. For the illustration, this ratio is 98% when $\alpha = 2$, 97% when $\alpha = 3$, and 93% when $\alpha = 10$, all for $n \geq 20$.

REFERENCES

ANSCOMBE, F. J. (1960) "Rejection of outliers" [with commentary]. *Technometrics* 2: 123-166.

ANSCOME, F. J. (1961) "Examination of residuals." *Proceedings of the Fourth Berkeley Symposium on Mathematical Statistics and Probability* 1: 1-36.

ANSCOMBE, F. J. (1973) "Graphs in statistical analysis." *American Statistician* 27: 17-22.

ANSCOMBE, F. J., and TUKEY, J. W. (1963) "The examination and analysis of residuals." *Technometrics* 5: 141-160.

ATKINSON, A. C. (1985) *Plots, Transformations, and Regression: An Introduction to Graphical Methods of Diagnostic Regression Analysis*. Oxford: Clarendon.

BARTLETT, M. S. (1937) "Properties of sufficiency and statistical tests." *Proceedings of the Royal Society* A 160: 268-282.

BECKMAN, R. J., and COOK, R. D. (1983) "Outliers." *Technometrics* 25: 119-163.

BELSLEY, D. A. (1984) "Demeaning condition diagnostics through centering" [with commentary]. *American Statistician* 38: 73-93.

BELSLEY, D. A., KUH, E., and WELSCH, R. E. (1980) *Regression Diagnostics: Identifying Influential Data and Sources of Collinearity*. New York: John Wiley.

BOX, G. E. P., and COX, D. R. (1964) "An analysis of transformations." *Journal of the Royal Statistical Society, Series B* 26: 211-252.

BOX, G. E. P., and TIDWELL, P. W. (1962) "Transformation of the independent variables." *Technometrics* 4: 531-550.

BREUSCH, T. S., and PAGAN, A. R. (1979) "A simple test for heteroscedasticity and random coefficient variation." *Econometrica* 47: 1287-1294.

CHAMBERS, J. M., CLEVELAND, W. S., KLEINER, B. and TUKEY, P. A. (1983) *Graphical Methods for Data Analysis*. Belmont, CA: Wadsworth.

CHATTERJEE, S., and HADI, A. S. (1988) *Sensitivity Analysis in Linear Regression*. New York: John Wiley.

CHATTERJEE, S., and PRICE, B. (1977) *Regression Analysis by Example*. New York: John Wiley.

CLEVELAND, W. S. (1985) *The Elements of Graphing Data*. Belmont, CA: Wadsworth.

CONOVER, W. J., JOHNSON, M. E., and JOHNSON, M. M. (1981) "A comparative study of tests for homogeneity of variances, with applications to the outer continental shelf bidding data." *Technometrics* 23: 351-361.

COOK, R. D. (1977) "Detection of influential observations in linear regression." *Technometrics* 19: 15-18.

COOK, R. D., and WEISBERG, S. (1982a) "Criticism in regression," in S. Leinhardt (ed.) *Sociological Methodology* (pp. 313-361). San Francisco: Jossey-Bass.

COOK, R. D., and WEISBERG, S. (1982b) *Residuals and Influence in Regression*. London: Chapman and Hall.

COOK, R. D., and WEISBERG, S. (1983) "Diagnostics for heteroscedasticity in regression." *Biometrika* 70: 1-10.

DANIEL, C., and WOOD, F. S. (1980) *Fitting Equations to Data* (2nd ed.). New York: John Wiley.

DAVIS, C. (1990) "Body image and weight preoccupation: A comparison between exercising and non-exercising women." *Appetite* 15: 13-21.

DIACONIS, P., and EFRON, B. (1983) "Computer intensive methods in statistics." *Scientific American* 248(5): 116-130.

DRAPER, N. R., and SMITH, H. (1981) *Applied Regression Analysis* (2nd ed.). New York: John Wiley.

DUNCAN, O. D. (1961) "A socioeconomic index for all occupations," in A. J. Reiss, Jr., with O. D. Duncan, P. K. Hatt, and C. C. North *Occupations and Social Status* (pp. 109-138). New York: Free Press.

ERICKSEN, E. P., KADANE, J. B., and TUKEY, J. W. (1989) "Adjusting the 1980 Census of Population and Housing." *Journal of the American Statistical Association* 84: 927-944.

FOX, J. (1984) *Linear Statistical Models and Related Methods.* New York: John Wiley.

FOX, J. (1990) "Describing univariate distributions," in J. Fox and J. S. Long (eds.) *Modern Methods of Data Analysis* (pp. 58-125). Newbury Park, CA: Sage.

FOX, J., and LONG, J. S. (eds.) (1990) *Modern Methods of Data Analysis.* Newbury Park, CA: Sage.

FOX, J., and MONETTE, G. (in press) "Generalized collinearity diagnostics." *Journal of the American Statistical Association.*

FOX, J., and SUSCHNIGG, C. (1989) "A note on gender and the prestige of occupations." *Canadian Journal of Sociology* 14: 353-360.

GALLANT, A. R. (1975) "Nonlinear regression." *American Statistician* 29: 73-81.

HOAGLIN, D. C., MOSTELLER, F., and TUKEY, J. W. (eds.) (1983) *Understanding Robust and Exploratory Data Analysis.* New York: John Wiley.

HOAGLIN, D. C., MOSTELLER, F., and TUKEY, J. W. (eds.) (1985) *Exploring Data Tables, Trends, and Shapes.* New York: John Wiley.

HOAGLIN, D. C., and WELSCH, R. E. (1978) "The hat matrix in regression and ANOVA." *American Statistician* 32: 17-22.

HOERL, A. E., and KENNARD, R. W. (1970a) "Ridge regression: Applications to nonorthogonal problems." *Technometrics* 12: 69-82.

HOERL, A. E., and KENNARD, R. W. (1970b) "Ridge regression: Biased estimation for nonorthogonal problems." *Technometrics* 12: 55-67.

KISH, L. (1965) *Survey Sampling.* New York: John Wiley.

KMENTA, J. (1986) *Elements of Econometrics* (2nd ed.). New York: Macmillan.

LEAMER, E. E. (1978) *Specification Searches: Ad Hoc Inference with Nonexperimental Data.* New York: John Wiley.

LITTLE, R. J. A., and RUBIN, D. B. (1990) "The analysis of social science data with missing values," in J. Fox and J. S. Long (eds.) *Modern Methods of Data Analysis* (pp. 374-409). Newbury Park, CA: Sage.

MALLOWS, C. L. (1973) "Some comments on C_p." *Technometrics* 15: 661-676.

MALLOWS, C. L. (1986) "Augmented partial residuals." *Technometrics* 28: 313-319.

MONETTE, G. (1990) "Geometry of multiple regression and interactive 3-D graphics," in J. Fox and J. S. Long (eds.) *Modern Methods of Data Analysis* (pp. 209-256). Newbury Park, CA: Sage.

MOSTELLER, F., and TUKEY, J. W. (1977) *Data Analysis and Regression: A Second Course in Statistics.* Reading, MA: Addison-Wesley.

NATIONAL OPINION RESEARCH CENTER. (1989) *General Social Survey* [data set]. Chicago: Author.

ORNSTEIN, M. D. (1976) "The boards and executives of the largest Canadian corporations: Size, composition, and interlocks." *Canadian Journal of Sociology* 1: 411-437.

OSTROM, C. W., Jr. (1990) *Time Series Analysis: Regression Techniques* (2nd ed.). Newbury Park, CA: Sage.

PINEO, P. C., and PORTER, J. (1967) "Occupational prestige in Canada." *Canadian Review of Sociology and Anthropology* 4: 24-40.

ROUSSEEUW, P. J., and LEROY, A. M. (1987) *Robust Regression and Outlier Detection.* New York: John Wiley.

SILVERMAN, B. W. (1986) *Density Estimation for Statistics and Data Analysis*. London: Chapman and Hall.

STATISTICS CANADA. (1971) *Census of Canada* (Vol. 3). Ottowa, Canada: Author.

STINE, R. (1990) "An introduction to bootstrap methods: Examples and ideas," in J. Fox and J. S. Long (eds.) *Modern Methods of Data Analysis* (pp. 325-373). Newbury Park, CA: Sage.

THEIL, H. (1971) *Principles of Econometrics*. New York: John Wiley.

TOBIN, J. (1958) "Estimation of relationships for limited dependent variables." *Econometrica* 26: 24-36.

TUKEY, J. W. (1977) *Exploratory Data Analysis*. Reading, MA: Addison-Wesley.

VELLEMAN, P. F., and HOAGLIN, D. C. (1981) *Applications, Basics, and Computing of Exploratory Data Analysis*. Boston: Duxbury.

VELLEMAN, P. F., and WELSCH, R. E. (1981) "Efficient computing of regression diagnostics." *American Statistician* 35: 234-241.

WEISBERG, S. (1985) *Applied Linear Regression* (2nd ed.). New York: John Wiley.

WHITE, H. (1980) "A heteroscedasticity-consistent covariance matrix estimator and a direct test for heteroscedasticity." *Econometrica* 38: 817-838.

WONNACOTT, T. H., and WONNACOTT, R. J. (1990) *Introductory Statistics* (5th ed.). New York: John Wiley.

ABOUT THE AUTHOR

JOHN FOX is Professor of Sociology and of mathematics and statistics at York University in Toronto, Canada, where he is coordinator of the Statistical Consulting Service at the Institute for Social Research. He earned a Ph.D. in Sociology from the University of Michigan in 1972. He has taught many workshops on statistical topics, at such places as the summer program of the Inter-University Consortium for Political and Social Research and at the annual meetings of the American Sociological Association. His recent and current work includes research on statistical methods and on the political economy of Canada. He is the author of *Linear Statistical Models and Related Methods* (Wiley, 1984), and co-editor (with Scott Long) of *Modern Methods of Data Analysis* (Sage, 1990).